The VINTAGE VIDEO BASEMENT

An Hachette UK Company
www.hachette.co.uk

First published in Great Britain in 2026 by Ilex Press, an imprint of Octopus Publishing Group Ltd
Carmelite House
50 Victoria Embankment
London EC4Y 0DZ
www.octopusbooks.co.uk

The authorized representative in the EEA is Hachette Ireland, 8 Castlecourt Centre, Dublin 15, D15 XTP3, Ireland (email: info@hbgi.ie)

Distributed in the US by Hachette Book Group, 1290 Avenue of the Americas, 4th and 5th Floors, New York, NY 10104

Distributed in Canada by Canadian Manda Group, 664 Annette St., Toronto, Ontario, Canada M6S 2C8

Commissioning Editor: Stephanie Selçuk-Frank
Art Director: Ben Gardiner
Designer: Chris Robinson
Senior Editor: Alex Stetter
Senior Production Manager: Peter Hunt

ISBN: 978-1-84601-615-8
eISBN: 978-1-84601-616-5

A CIP catalogue record for this book is available from the British Library.

Printed and bound in China.

10 9 8 7 6 5 4 3 2 1

MIX
Paper | Supporting responsible forestry
FSC® C008047

The VINTAGE VIDEO BASEMENT

ICONIC
ELECTRONICS
FROM THE
70s, 80s & 90s

MATT PARKER

ilex

CONTENTS

WELCOME TO THE BASEMENT

Come on in! I'm Matt Parker, and I am an avid collector and preserver of vintage physical media and electronics. You may know me as the person who runs the *Vintage Video Basement* account that is popular across multiple social media platforms, including Instagram, TikTok, YouTube, and Facebook. It is my mission to take advantage of modern technology to share my passion for vintage electronics and nostalgic memorabilia—in particular home video equipment from the early days of the home theater, such as CRTs, VCRs, and video cameras. I repair and collect these devices to showcase them in my basement with other décor that fits their style and the time period in which they were produced.

I have always been a collector. Over the years, my collections have ranged from *Star Wars* memorabilia to elements of the periodic table. I inherited my enthusiasm for collecting from my dad. He always collected antiques, and would bring home items he had found. In fact, the first Betamax I ever owned, a 1983 Sony SL-2300, was given to my dad, and he passed it on to me. Having only owned VHS machines up to that point, the Betamax sparked my fascination with the history of the format wars, and how multiple physical media formats battled it out before VHS became ubiquitous.

This resulted in my VHS collection bleeding over into other formats (Beta, LaserDisc, CED videodisc ...) and I realized that I enjoyed the VCRs and players as much as the movies. Different machines have their own design and aesthetic, which makes viewing classic movies a more enjoyable experience. This is further enhanced by hooking the machines up to CRT (cathode-ray tube) televisions from the same era—those heavy old TVs that are often as deep as they are wide. Once I had my own space in which I could display my collection, the *Vintage Video Basement* emerged.

After decades of collecting, I had accumulated a vast assortment of items, ranging from electronics and toys to furniture. The intent of the *Vintage Video Basement* is to create a space that encapsulates the aesthetic I enjoy the most. It is not intended to represent any particular year. Instead, it is meant to capture the feeling of a time before the internet, when everything was analog. That bygone era when the electronics were loud, thanks to their clicks and thumps; when you could feel the static electricity on the TV screen; every cartoon had a toy line of the best figures and playsets ever made; we rode our bikes all day, and landed at a friend's house to watch a movie rented from the local video store; and it felt like every experience was exciting and new.

The main focus in the basement is electronics from the Seventies, Eighties, and Nineties. The TVs and VCRs many of us grew up with (and some you may have never seen) are what make the basement unique and interesting, but the wood paneling, shag carpet, and lighting really make it feel like you're stepping back in time. This book is intended to show off my collection, while discussing the importance of each piece and why I love every one. It will be a walk down memory lane, or it may make you nostalgic for a time you never experienced. So, make yourself at home and let's go back to a time when the electronics were analog, the toys were fun, the TVs were thick, and everything was wrapped in wood.

ZENITH
THE QUALITY GOES IN BEFORE THE NAME GOES ON.
WELCOME TO THE VINTAGE VIDEO BASEMENT
PEPSI
· VHS · BETAMAX · LASERDISC · BLU-RAY ·
NOW SHOWIN
MY BLOODY VALENTINE & MAXIMUM OVERDRIVE
THE GOONIES
JVC
HALLOWEEN
JAWS
CHRISTINE
SONY
BLOCKBUSTER VIDEO

ITEMS OF INTEREST 01

1 TOYS 'R' US SIGN LETTER

This iconic backwards 'R' was saved from a defunct Toys 'R' Us in the Seattle, Washington, area. I added lights and mounted it on the ceiling to create the most nostalgic lighting.

2 1981 SONY TRINITRON KV-2145R

This 21-in Trinitron is the perfect size for any occasion. The screen is large enough to produce a clear picture that is visible from anywhere in the room, while the set is light enough to be easily moved if you want to rearrange the space.

3 WARM LIGHTING

Notice the warm lighting from the tension pole lamp. This gives the basement its vibe and is what really sets the mood of the photo.

4 1975 SONY TRINITRON LV-1901

Although the LV-1901 (see page 72) does work, it is not turned on in this picture because, unlike the other TVs, it does not display static. The screen only shows a picture when a signal is detected.

5 JACK-O'-LANTERN

Halloween is a big deal in the basement, and I always go heavy with the decorations throughout the fall.

1
2
3
4
5
ZENITH
THE QUALITY GOES IN BEFORE THE NAME GOES ON.
WELCOME TO THE VINTAGE VIDEO BASEMENT
PEPSI
· VHS ·
· BETAMAX ·
· LASERDISC ·
· BLU-RAY ·
OUIJA
GAME BOY
Barrel of Fun
TONY HAWK
BLOCKBUSTER VIDEO

As we enter the basement, let's discuss that cozy feeling you will encounter as you flip through the following pages. Yes, I'm referring to nostalgia. As we all know, nostalgia is that feeling you get when confronted with a sensation (smell, sound, taste, etc.) from a previous experience in your life. This feeling is usually a happy or bittersweet longing for a time you know will never exist again, and it can often remind you of an activity, friendship, or item you had long forgotten. Nostalgia can make you feel calm and cozy, or sad and lonely.

People can, however, have nostalgia for a time they never experienced. Songs, automobiles, and even electronics can transport people to a time before they were born. The term "anemoia" was coined by John Koenig in *The Dictionary of Obscure Sorrows* to describe this unusual type of nostalgia. Typically, anemoia is brought about by movies, books, or music. Many movies from our childhood are set in a time that the adults who made them were nostalgic for. For example, *Back to the Future* came out in 1985, but much of this movie was set in 1955. *The Sandlot* came out in 1993, but takes place in 1962. As we began to leave the Nineties and enter the Noughties, *That Seventies Show* was a popular on TV. In the 2010s, we got *Stranger Things*, which of course takes place in the Eighties. Even now you can see a trend for entertainment reflecting the Nineties becoming popular. Every generation idealizes the time of its youth and introduces it to younger generations. This in turn can give young people anemoia for a time that has been glamorized and wiped clean of many of its hardships and struggles. Those time periods are looked at through "rose-colored glasses," as the saying goes, which doesn't capture the true feeling of living through them. I actually remember living through the Nineties and thinking it was much drabber and more depressing than the way earlier decades were portrayed in the movies I watched. I longed for a time I had only experienced through the TV. Now I look back and think about how awesome the Nineties were. Time has washed clean the aspects of that decade I didn't enjoy, and nostalgia has curated and highlighted all the best parts. As a result, I would love to go back to the Nineties.

People often think they are living through the worst period of history, and, if they could just go back to a different era, everything would be much better. I am more of the mind that we should better understand the past in order to appreciate how far we have come. It's important we experience the present while also keeping the aspects of the past we most enjoy alive. There is some joy to be had in looking back at the past and marveling at how large, analog, and mechanical many daily devices used to be. We've come a long way, technologically speaking, but that doesn't mean these obsolete machines from the past can't still be enjoyed. I intend for the entries that follow to bring you comfort in some sense, and whether it is nostalgia or anemoia you feel, understand that we are looking at the past through a lens of coziness and convenience. That lens, however, is more like a tube—a cathode ray tube—and we are just watching it.

MASTERS
OF THE UNIVERSE
SUPER MARIO BROS./DUCK HUNT
Nintendo
ENTERTAINMENT SYSTEM
SONY
FRIDAY THE 13TH
FRIDAY THE 13TH PART 2
FRIDAY THE 13TH PART 3
FRIDAY THE 13TH PART V
BLOCKBUSTER VIDEO

01

DON'T SIT TOO CLOSE TO THE TV

The television was introduced to the world by the Radio Corporation of America (RCA) at the New York World's Fair in 1939. This single device went on to shape the fabric of our society. The cathode-ray tubes (CRTs) in original TVs were so long that the screen faced the ceiling, and the image was reflected back at the viewer using an angled mirror. The sizes ranged from 5 to 12in, with prices between $199.50 and $600. Of course, the early TVs only produced a black-and-white image. By 1950, close to 15 million black-and-white televisions had been produced, but a new breakthrough was on the horizon. In 1953, RCA developed an all-electronic system for the reception of color images that would not interfere with the current black-and-white reception. This was adopted by the National Television System Committee (NTSC), and the color TV was born.

My collection consists primarily of color TVs. Many of the early black-and-white models are absolutely stunning, but I prefer to use sets that fit within the timeframe of the rest of the items I collect, and for me there is nothing that tips the nostalgia scale as quickly as a wood-grain tube TV with chrome accents and big tuning knobs. In the right setting, one of these TVs is essentially a portal to a simpler time when everyday electronics were electromechanical marvels.

My favorite eras for CRTs are the Seventies and Eighties, because that was when color TV was hitting its stride and companies were pushing the technology to be bigger and better. One of the organizations leading this push was Sony, and as you flip through my collection you will see that many of my TVs are Sony Trinitrons. Sony revolutionized the color television when it introduced the Trinitron in 1968. Until then, the typical color television consisted of three electron guns. Each gun shot one primary color (red, green or blue) through a small lens. After that, the colour passed through a shadow mask and struck the viewing surface of the TV. The shadow mask was an array of holes in a screen that was intended to only allow one color beam through each hole, while the other color beams would be blocked at that location. Back this idea out full scale, and it is clear a majority of the electrons could never make it to the screen, which meant the image was not as bright as it could have been.

Sony is always working to revolutionize tech, so it introduced a single electron gun that shot all three colors through a single, large lens. These electrons then passed through an aperture grille. The aperture grille consisted of slits across the screen that separated the colors but allowed a majority of the electrons to hit the screen. This technological improvement allowed for a brighter and sharper image, plus it reduced the need for continuous adjustments. It was so extraordinary that in 1973, Sony won an Emmy for its work in improving color television.

Now, don't get me wrong, I do not exclusively collect Sony. I have some great offerings from companies such as Zenith, RCA, and Panasonic. Anything I come across that is cool or unique can absolutely find its way into the basement. So let's ignore the title of this chapter and start looking at these incredible television sets. Feel free to sit as close to these TVs as you like. It will provide a sense of comfort you have not felt in years.

1969 SONY TRINITRON KV-1210U

Screen size: 12in **Weight:** 36lb 2oz (16.4kg) **Introductory price:** $319.95 USD

Let's start with the first Trinitron released in the United States. After several years of research and development, Sony introduced the Trinitron in Japan in 1968, and in the United States the following year. Some of the first programs you could watch on a newly acquired Trinitron included the Apollo 11 Moon landing, when Neil Armstrong became the first man to walk on the Moon, or maybe even the premiere of *The Brady Bunch*, when Mike and Carol got married and the world's favorite family of eight was formed. In an effort to release the Trinitron as quickly as possible, Sony only made it available as 7-in and 12-in models. Additional development was needed before larger screens were introduced, and Sony could not delay the release of the Trinitron any longer. Quick side note: when discussing the screen size of televisions, it will always be the diagonal measurement and not the height or width. Even the screen size of modern televisions is measured diagonally, from one corner to another.

The KV-1210U was the 12-in version wrapped in a walnut wood housing. Even though it weighs close to 40lb, the set was considered to be portable and had a spring-action handle on the top of the case. The iconic Trinitron logo, with its red, green, and blue ovals, is featured prominently under the speaker. The KV-1210U really demanded attention, thanks not only to its beautiful design but also to its incredibly sharp and vivid image. A key identifier of the Trinitron is the cylindrical screen, which only curves horizontally from one side to the other and is flat along the vertical plane, unlike the screens of other TVs of the time. This resulted in less distortion at the corners. Thanks to the one-gun system, aperture grille, and cylindrical screen combination, it was truly a dramatic step forward for color television technology.

WHY I LOVE IT

It is the first Trinitron! It was a game-changer for color TV and helped establish Sony as the pinnacle manufacturer for both consumer and professional electronics. Also, it is just a beautiful and sleek design.

TRINITRON
SONY
SOLID STATE
TRINITRON COLOR
PULL ON/VOL

1971 SONY TRINITRON KV-1212

Screen size: 12in **Weight:** 31lb 5oz (14.2kg) **Introductory price:** $369.95 USD

In 1971, while shows such as *All in the Family* and *Columbo* dominated the airwaves, Sony followed up the KV-1210U with the KV-1212. Although the style was similar, the new look was more refined. The common controls no longer extend under the screen, but are to its right from the viewer's perspective. The power is engaged by pulling the volume knob, and there is even a button for automatic fine tuning (AFT) that controls the color saturation and hue while fine-tuning the picture and sound reception. The color and picture controls can also be adjusted manually. The handle on top was modified to simply lay down when not needed, and, at just over 31lb, this set can be moved from room to room with ease.

"The introduction of the Sony Trinitron was a game-changer for color TV."

WHY I LOVE IT

This is one of the early Trinitrons that is more popular with collectors. The wood grain and large tuner knobs really give it that iconic TV style. This one looks great in my basement kitchen next to my vintage coffee maker on the orange countertop.

1971 SONY TRINITRON KV-1710

Screen size: 17in **Weight:** 57lb (25.8kg) **Introductory price:** $469.95 USD

Sony's first step toward producing larger screens was this 17-in version of the Trinitron; a 15-in version was released soon after. These sizes were not even close to the largest color CRT available on the market at the time, but Sony was beginning the push to build bigger and better Trinitrons. The style cues of the smaller sets were still present, such as the walnut-grain finish and black-and-chrome accents, as well as features such as automatic color and fine-tuning. Despite its weight, this television was still considered by Sony to be portable, and included a pop-up handle on the top. Although Sony ran an ad to promote the new 17-in screens that showed a man lying in bed with one on his torso, the company also claimed this was not a "tummy TV," which is a good thing. At close to 60lb, it's best to keep this one on the dresser.

"Nothing tips the nostalgia scale as quickly as a wood-grain tube TV with chrome accents and big tuning knobs."

WHY I LOVE IT

This 17-in Trinitron was Sony's initial push to show how large it could go with the Trinitron technology—and it was just getting started. I particularly like how this set looks on the rolling TV stand.

AUTO/AFT
SONY
TRINITRON
SOLID STATE

1972 SONY TRINITRON KV-1730R

Screen size: 17in **Weight:** 62lb (28.1kg) **Introductory price:** $549.95 USD

The first Sony Trinitron with its own remote control came just four years after the model was introduced, but the remote was only available on the higher-end KV-1730R model. You can determine that this was the 17-in version—the largest screen size available in the Sony line-up at the time—from the model number. You'll also notice the "R" at the end of the model number, which identifies it as a remote-controlled unit. What's cool about this TV is that it had the same big tuning knobs as the rest of the Trinitron lineup, and the remote control was ultrasonic, meaning it used high-pitched sounds that could not be heard by the human ear to control the TV. If the UHF or VHF buttons were pressed to prompt the channel to change, a solenoid was activated in the TV to turn the corresponding tuning knob, resulting in a loud bang for every channel that went by. It is quite dramatic and something to experience. You could also only change the channel in the forward direction—there was no going back to a lower channel without circling the entire dial. Luckily, there were only 13 channels at the time, if you stuck with VHF. Due to the loud "ka-ching" of the tuning knob as it turned, you wouldn't want to do this with a family member sleeping nearby. To ensure the remote control wouldn't cause such a disturbance, the brightness control knob on the rear of the TV could be pulled out to disable the remote. This also prevented any stray noises at the same frequency as the remote from altering the TV unexpectedly. The ultrasonic remote wasn't anything new in the industry. Companies such as Zenith and RCA had been using them for years, but this was one of the rare examples of a Sony ultrasonic remote. Within a few short years, Sony would move to the more advanced infrared remote control.

It's also worth noting that there was a holster on the back of the TV for the remote, so it would not go missing. Additionally, the TV had a swivel base allowing for easy adjustment of the viewing direction. Even though the KV-1730R was just a few pounds heavier than the KV-1710 (see page 18), Sony eliminated the single pull-out handle on top in favor of grip handles on either side of the housing, which is more convenient for carrying.

WHY I LOVE IT

This TV was given to me when I purchased the 1981 Sony KV-3000R, which you'll read about on page 26. The former owner had left it in the garage. I later discovered they had been the Vice President of Sales for Sony during the Seventies, Eighties, and Nineties. It's just such an amazing piece of tech and a cool piece of the Sony Trinitron's history.

TRINITRON
VHF
UHF
SOLID STATE
SONY

SONY
ON/OFF

SONY
ON/OFF
VOL
UHF
VHF

ITEMS OF INTEREST 02

1

1977 FISHER MC-3160 INTEGRATED STEREO SYSTEM

This stereo system from the late Seventies incorporated a record player, cassette player, and 8-track player along with an AM/FM tuner into one compact system. I love how audiophiles were able to enjoy several formats within the same device.

2

1976 SONY TRINITRON KV-1723D & 1976 SONY TRINITRON KV-1215

I found both of these sets at an estate sale outside of Portland, Oregon. The former owner collected vintage electronics and had multiple rooms that were stacked floor-to-ceiling with audio and video equipment. I was able to dig out several Beta players and repair parts as well.

3

ORIGINAL ZENITH ADVERTISEMENT CLOCK

This light-up clock would have been found in a Zenith dealer shop throughout the Seventies and Eighties. It's an eye-catching piece designed to ensure customers would consider this American brand when choosing electronics. I found this one in the backroom of an antique store.

4

SHAG CARPET

One of the pieces of décor that sets the basement apart is the shag carpet. This type of flooring is rarely seen these days, but it really pulls the electronics, wood paneling, and other vintage pieces together.

WELCOME TO THE VINTAGE VIDEO BASEMENT
PEPSI
· VHS · BETAMAX · LASERDISC · BLU-RAY ·
ZENITH
THE QUALITY GOES IN BEFORE THE NAME GOES ON
OUIJA
FRIDAY THE 13TH
TURTLES
SONY
E.T.
Barrel of Fun
BLOCKBUSTER VIDEO
1
2
3
4

1977 ZENITH PANORAMA VI (SJ2541X)

Screen size: 25in **Weight:** about 200lb (90.7kg) **Introductory price:** $790 USD

The Panorama VI is a part of Zenith's Chromacolor II line. The term Chromacolor II refers to the picture tube, which produces a sharp, high-quality color image. These solid-state sets were designed to be easy to service—failed modules could simply be replaced, eliminating the need for component-level troubleshooting, which requires considerably more time. Its curved housing made this set ideal for corner placement, and even makes the screen seem a bit larger than it actually is. The silver bezel with black accents gives the TV a futuristic feel. If you consider that 1977 was the year *Star Wars* was released, this set looks like it would fit right in on the Death Star. The rosewood veneer across the top and sides, however, places it back on Earth circa 1977. An interesting feature of this set is the ability to zoom in on the picture: the remote control has a button that makes the center of the image 50 percent larger. I'm not sure when anyone would have really needed that option, but it was there.

Another cool feature that came on later versions of this, and many other Zenith models, was the Space Phone. This allowed the user to hook their landline phone directly into the TV so that they could take calls on it without having to get out of their recliner. Society may have peaked when a phone call could be answered through the TV, but now people are watching TV on their phones, and that seems like a downgrade.

WHY I LOVE IT

The design of this TV is unmistakably Zenith. The manufactuerer made TVs in this style for many years, and I think they are beautiful. After searching for a couple years I ended up finding this one in Idaho while on a road trip. After a quick detour, the Zenith was loaded up in the back of my vehicle. I travel with large electronics more often than you might think, so it wasn't an issue.

ZENITH
THE QUALITY GOES IN
BEFORE THE NAME GOES ON.

1981 SONY TRINITRON KV-3000R

Screen size: 30in **Weight:** 510lb (231.3kg) **Introductory price:** $10,000 USD

"With a whopping 30-in screen, this was the largest tube TV you could buy in 1981."

Imagine this—it's 1981 and you are in the market for the largest consumer color television ever built. You aren't interested in projection televisions. You want a big-screen picture tube with an incredible sound system. Good news: Sony has just released such a TV. Bad news: it costs $10,000!

That's right. In 1981, Sony went all-in on the most luxurious Trinitron ever built—the KV-3000R. With a whopping 30-in screen, this was the largest tube TV you could buy at the time. Sony was among the companies pushing CRT technology to the largest size possible, and it had come a long way from its initial 12-in tube. But this wasn't just a TV, it was an entire system. It included a pair of massive speakers capable of resonant 3D-stereo with matrix sound. The system had motorized tambour doors that could be controlled with the push of a button from the Express Commander remote control. The console was wrapped in rare tamo ash wood from a remote forest in Hokkaido, Japan. This wasn't your grandma's wood console Magnavox with the doily and glass dish of potpourri on top. This was an extreme display of wealth and style.

The Trinitron was more than just an entertainment center; it was a statement piece designed to show off how wealthy the owner was. As it was limited to just 1,000 units, owners had the option of getting their name and production number engraved on a special badge placed next to the power button on the console. The KV-3000R was available in the 1981 Neiman Marcus catalog for the extortionate price of $10,000, and came with a matching Betamax stand. It may seem like it was an outlandish price, and it probably was, but that 30-in screen was cutting-edge technology at the time. The CRT could also be purchased as a professional monitor, the CVM-3000, which cost $5,680 but did not have the lavish housing and luxurious features.

JAWS
The night no one comes home.
SONY
SONY

JAWS
The night no one comes home.
SONY

I also believe Sony priced this unit at such a high price as a ploy to sell its more affordable products. One ad reads, "If you can't own our $10,000 console, here's some consolation," and has a picture of several other Trinitron sets. It goes on to describe how all the Trinitrons are worth $10,000, but they are available at a much more reasonable price. Pretty genius.

Another cool fact is that a KV-3000R was featured on no less than five episodes of *The Price is Right*. It would always be a feature in the Showcase, and the contestant doing the bidding was sure to bid way too low. I mean who would have expected a $10,000 TV? That's more than many of the cars Bob Barker was giving away on *The Price is Right* at the time! But don't worry. You got free delivery and professional installation by Sony technicians included in the price. And as the KV-3000R weighed in at a massive 510lb, that was a necessity.

WHY I LOVE IT

This is an all-time great TV and exactly the kind of thing I enjoy most. It is very unusual and has some amazing features such as the automatic doors. It was limited to just 1,000 units, and mine is #458. This one came up for sale in the Upper Peninsula of Michigan and had all the things I was hoping to find when searching for one: the matching Betamax stand (see photo opposite), remote controls, schematics, operating instructions, and even extra pieces of tamo ash wood veneer in case the housing became damaged. Being impatient, I rented a trailer and made the trip up to Michigan in the winter. We had just missed a snowstorm and the roads were icy, but nothing was going to prevent me from finally owning a KV-3000R. After safely returning to the basement, I did some investigating on the original owner and discovered they had been the VP of Sales for Sony in consumer electronics during the release of several big Sony products including the Betamax, Walkman, and CD. I wish more of my electronics had stories this good.

SONY
REMOTE COMMANDER
RM-706W
PICTURE
CHANNEL
CLEAR
MUTING
VOLUME
RECALL
POWER
SONY
REMOTE COMMANDER
RM-706

SONY
SONY

1981 PANASONIC CT-9071

Screen size: 19in. **Weight:** 56lb 11oz (25.4kg) **Introductory price:** about $600 USD

The CT-9071 was part of Panasonic's ColorPilot Series II line, which means it has a "ColorPilot" button that could be pressed to trigger it to constantly self-adjust the color for optimal viewing pleasure. Or at least that is how it was marketed. The overall design of the TV is classic, with silver accents that provide a contrast to the brown tones and wood veneer. For me, the real standout feature of this set is that the controls on the front are actually a remote control that is magnetically attached to a spring-loaded panel that pops out with the push of a button. Back in 1981, MTV was a new channel and its marketing campaign had everybody saying, "I want my MTV!" The owner of this Panasonic could remove the remote and take it back to the sofa to watch music videos at the right volume.

"Remember when MTV played music videos? Those were the days."

WHY I LOVE IT

I came across this Panasonic in a listing from a guy who was trying to get rid of several CRTs that were stored in his garage. He had just bought a new car and needed to clear some space so it would fit. I ended up purchasing this Panasonic and three Trinitrons that day. The pop-out remote control was cool enough for me, and so it had to come back to the basement.

1981 SONY TRINITRON KV-2645RS

Screen size: 26in **Weight:** about 210lb (95.2kg) **Introductory price:** about $1,100 USD

If you're a fan of early-Eighties horror, you may recognize this floor model Trinitron. One was featured prominently in the 1982 Tobe Hooper film *Poltergeist*. It even made the poster and cover art for the movie with the little girl, Carol Anne, sitting in front of the TV with her hands pressed against the static-filled screen. Could you imagine trying to make a scene like that in the modern era, with TVs mounted so high on the wall? Can a modern flatscreen even be scary? I don't think so. This sort of horror iconography is only possible with classic CRTs at ground level. There were several other TVs in the movie, and even other Trinitrons, but the KV-2645RS gets the most attention. Its placement in the movie has made it one of the more iconic Trinitrons ever made.

The 26-in screen was considered giant in 1981. As we saw with the KV-3000R (see page 26), the 30-in screen was the largest direct-view TV that had been produced at the time and was priced far out of range for almost everyone. The 26-in screen was close enough in size for just one-tenth of the price.

The design of the console is basic but refined, with black and silver features that pop against the oak wood case. The silver mesh covering on the speakers has just the right amount of shimmer and gives the set a hint of elegance. The KV-2645RS also has the matrix sound feature, which simulated stereo sound before it was prevalent in broadcasting or home video.

WHY I LOVE IT

As a horror fan, of course I love this one—it's the model used in the classic horror film *Poltergeist*. And yes, I have recreated the poster from the movie with myself in front of the TV instead of Carol Anne. Oddly enough, I picked up this set for free on the road trip when I collected the KV-3000R.

"THE
TEXAS
HAINSAW MASSACRE"
America's most
bizarre and
brutal crimes!...
happened is true. Now the motion picture that's just as real.
E.T.
SONY
SONY

ITEMS OF INTEREST 03

1 **OAK ENTERTAINMENT CENTER**

These behemoth entertainment centers were a common sight in many homes of the Nineties, and were ideal for displaying audio and video equipment along with the TV. This one currently houses a VHS VCR, two Beta players, two LaserDisc players, and a CED player. The bottom compartments are packed with several formats of physical media.

2 **1987 ZENITH SYSTEM 3**

This 25-in Zenith is the perfect fit for the entertainment center, and is a great display for viewing the format of your choice. In fact, you can select your preferred format player using the switchboard hidden in the bottom of the center.

E.T.
NERF
ArrowStorm
NIGHT OF THE CREEPS
2
SONY
ASTEROIDS
SLOT RACERS
1
BLOCKBUSTER
VIDEO

1981 SONY TRINITRON KX-2501

Screen size: 25in **Weight:** 121lb 4oz (55kg) **Introductory price:** $1,300 USD

For years, the television had been a plug-and-play system, easily operated by both the videophile and the common man, but the early Eighties introduced a new kind of TV. The component TV gave consumers more flexibility and the capability of building a hi-fi system comprising individual parts. Instead of buying a console TV complete with tuner, speakers, and limited inputs/outputs, customers could purchase a stripped-down monitor giving them the freedom to select the speakers, tuner, and other components they preferred.

To accommodate this market, Sony introduced its Profeel line with 19-in and 25-in versions of its TV monitor. I have the 25-in Profeel, which has 350 lines of resolution. It was a significant improvement over the 280 lines of resolution used by conventional sets of the same size at the time. A sheet of contrast glass was attached to the front of the tube to improve contrast and prevent glare from the curved screen. Aesthetically, this also gave the Trinitron a cleaner look. In addition, RGB inputs made this Trinitron ideal for computer graphics.

Two types of speakers were offered—the side-mounted SS-X1A speakers and the larger SS-X10A bookshelf speakers. You could also, of course, choose other speakers if desired. If you wanted to use the monitor to view TV channels, you needed a tuner, as one was not built into the monitor. Sony offered the VTX-100R Access Tuner to allow for access to VHF, UHF, and cable channels. The tuner also allowed you to incorporate your video player(s) and sound system, while having everything controlled by a single remote.

> "At a time when most TVs were ornate and wrapped in wood veneer, the Profeel system was metallic and sleek."

ON VIDEOCASSETTE
TRINITRON
SONY
ANALOG/DIGITAL RGB
SONY
SONY
Betamax
SONY

Although the Profeel systems gave videophiles more freedom to build their dream machines, component TVs did not become a common sight in the average home. Consumers tend to like things that are simple and convenient. At a time when most TVs were ornate, wrapped in wood veneer, and had many visible controls, the Profeel system was metallic, sleek, and low profile. It was a high-tech precursor to the future design of CRTs before the transition to plasma and LCD TVs.

WHY I LOVE IT

My wife found this Profeel monitor at an estate sale in Sammamish, Washington. It was buried under a table in the garage. She sent me a photo asking if I was interested and told me they had priced it at $50. Of course I was interested, but I knew that the following day the estate sale would have everything marked 50 percent off, and I was willing to risk that no one else would be eyeing it before I could get there. I showed up the next day and there it was, still sitting under the table in the corner of the garage. I managed to pull it out and load it into my truck for a mere $25. I later found the SS-X10A bookshelf speakers in a barn that had been used for storage and was being cleaned out. As I was digging through the heaps of junk spread across the floor, I noticed those silver, tower speakers and knew what they were immediately. I bought them and an assorted bundle of other finds for just $20. You never know what you will find at a sale unless you look!

STEREO
ON VIDEOCASSETTE
SONY
SONY
SONY
SONY
Betamax

1982 RCA COLORTRAK 2000

Screen size: 25in **Weight:** about 250lb (113.3kg)

Now this is one of my more unusual and unique televisions. I've found almost no information about this TV from RCA. It is model number GGR2090A, or at least that's what the label says. For the longest time, the only way I knew it was a Colortrak 2000 was from the markings on the remote control, but it doesn't say that anywhere on the TV. The only other photo I had seen of this model was a single RCA ad that shows it in the home of a wealthy family. That was until I finally came across a 1982 RCA brochure that showcased its entire television lineup for the year. The odd thing is that the brochure lists it as model number GFR1001R. I'm not sure why there is a discrepancy between the model numbers, but it is definitely the same TV. The brochure refers to it as "a unique avant garde showcase for RCA's finest Colortrak TV performance."

What's so unusual about this TV is that it has a window across the front that folds down for cleaning. The window is a sheet of contrast glass that is intended to increase the contrast of the screen while also reducing glare. The controls are on top of the TV, hidden behind a cover that slides out of the way like a pocket door. The cabinet is wrapped in acacia wood veneer, and besides one RCA badge on the bottom-left, which is hidden behind the contrast glass, there are no other markings that identify the brand or model. People often ask me if the TV is in a custom cabinet.

Some of the features include the Super AccuFilter 100° picture tube and a Dual Dimension Sound System. This TV came out just before stereo audio was available through broadcast TV, cable TV, or videocassettes. As it was becoming a reality, RCA ensured its top-of-the-line consoles were capable of stereo. The speakers on this particular set are hidden across the bottom of the console under the picture tube.

WHY I LOVE IT

I purchased this TV from the original owner in Tacoma, Washington, who reminisced about finding the set in a local TV shop, and how it required a couple of guys to deliver it to his house. I could tell he had really enjoyed this TV and was proud of owning it. I assured him it was going to a good home and would be well taken care of. When I first came across the TV on Facebook Marketplace, I had no idea it would turn out to be such an intriguing console. The photos in the listing were dark and blurry, but I could tell there was a contrast glass cover and a cool design, so I was interested. I still waited a couple of weeks hoping someone else would get it, but when that didn't happen, I decided it had to be mine. It turned out to be one of my more popular sets. The fold-down glass is just too cool a feature and really grabs people's attention.

SCREAM GREATS
FRANKENSTEIN
VIDEODISC IS HERE!
GHOSTBUSTERS
FRIGHT NIGHT
HALLOWEEN
STAND BY ME
ALIEN
ARACHNOPHOBIA
JURASSIC PARK
JACK FROST

COLORTRAK 2000

1984 SONY TRINITRON KV-2670R

Screen size: 26in **Weight:** 114lb 10oz (51.7kg) **Introductory price:** $945.95 USD

For those who wanted a "giant screen" without having a large console that sat on the ground, the "consolette," as Sony described it, was a solution. This 26-in CRT is more compact than the floor-model console and could be placed on a cart or in an entertainment center for those who preferred their TV to be closer to eye-level. It is wrapped in a beautiful dark Brazilian rosewood veneer and, while not capable of stereo sound alone, could be hooked to a multichannel sound adapter through the multiplex (MPX) output to make it compatible with stereo.

I found this pristine example in Bellevue, Washington, at a condo that was the owner's vacation home. According to the seller, the owner rarely visited the condo, and the TV had not seen much use. The set was built into a custom entertainment center that had helped protect it for several decades. Although the set needed a little work due to components that had aged poorly, I was able to repair it easily, and it is now one of my more frequently used TVs.

WHY I LOVE IT

This really massive TV is designed in such a "slim" fashion. At the time, most tubes of this size were destined for large consoles that sat at ground level. This set broke free from the floor and gave the viewer more options for placement and viewing height. As it weighs 114lb, however, it would have been nice if Sony had included handles to assist with relocation.

1984 SONY TRINITRON KV-2681R

Screen size: 26in **Weight:** 210 lb (95.2 kg) **Introductory price:** $1,299.95 USD

"This TV harkens back to the Space Age, when society was looking toward the future with hope and optimism."

As you can probably tell by now, I prefer to collect the larger console televisions from the early-Eighties. This was the peak era for TV design in my opinion. These consoles were pieces of furniture that offered as much as they could with the current technology. The KV-2681R was no exception. At a time when most sets were covered with wood grain, this Trinitron is painted space-age pearl white and sits atop a chrome base. It harkens back to the Space Age, when society was looking toward the future with hope and optimism.

I searched for this model for several years as I wanted to add a clean, white console that would stand apart from the many wood-grain sets I already had. This attention-grabbing set really pops against the browns and oranges that surround it in the basement. I was surprised to find this one at a vintage store in Buffalo, New York, which also happened to have the Advent VideoBeam 1000A (see page 50) and the Quasar Great Time Machine (see page 88) that we will discuss later.

WHY I LOVE IT

For the most part, I am not a fan of the TV sets from the Nineties that were made of black plastic and had no character, but a Trinitron from the Eighties coated in a pearlescent white finish with chrome accents is right up my alley.

WARNING!
THIS LUNCH
PROTECTED BY
My Pet
MONSTER
SONY
SONY

1989 PANASONIC AG-500R

Screen size: 10in **Weight:** 26lb (11.7kg) **Introductory price:** $1,250 USD

We're going to end this TV section with an unusual CRT that is also popular with the collecting community—the AG-500R. This Panasonic is often referred to as "the microwave," due to its resemblance to the kitchen appliance. This familiar look comes from the protective cover that gives the CRT a flat appearance, along with the large button controls next to the screen. Its color scheme and small size also make it look as though it would fit well on the kitchen counter, and that just happens to be where I keep mine.

What really makes this set special is the built-in vertical VHS player. That's right. A VHS could be inserted vertically for playback only, because this machine is not capable of recording. Not only that, but the TV does not have a tuner and was, therefore, not able to pick up any channels, because this was not a consumer set. It was intended for educational and commercial situations. Imagine going to the library to do some research and needing to watch an informational VHS, or starting a new job and being required to watch a training video. You may have been given access to this set for those purposes. These days, the AG-500R is popular due to its unusual look, its portability, and its ability to play VHS. It was a precursor to the TV/VCR combos that became popular between the late Nineties and the early Noughties.

WHY I LOVE IT

Most TV/video player combos that came along in the following decades have the player under the TV, but the layout of the AG-500R can't be beat. The vertical player on the side of the picture tube is what makes this set special.

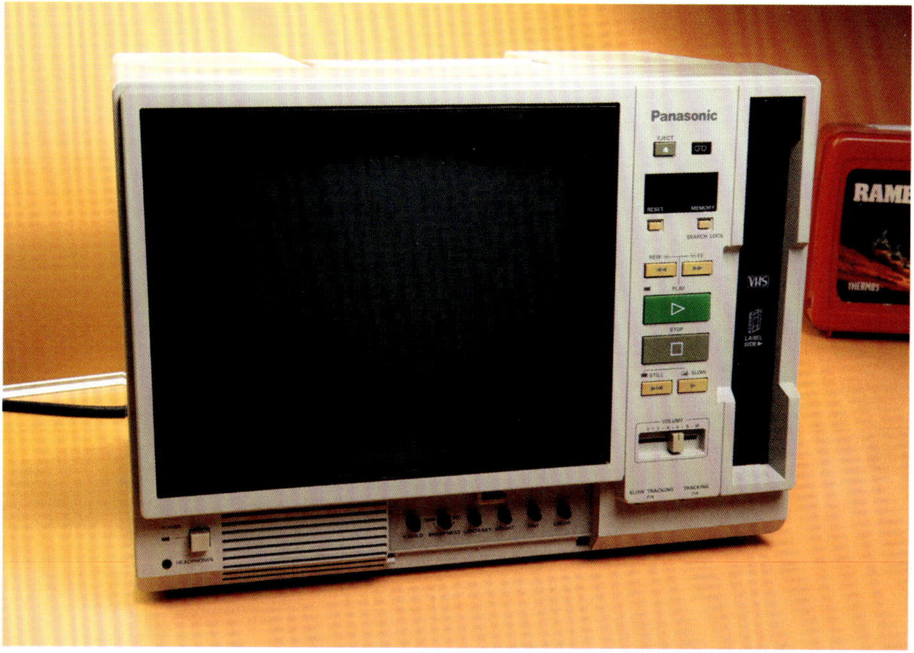
Panasonic
EJECT
RESET
MEMORY
SEARCH LOCK
PLAY
STOP
STILL
SLOW
VOLUME
VHS
LABEL SIDE
TRACKING

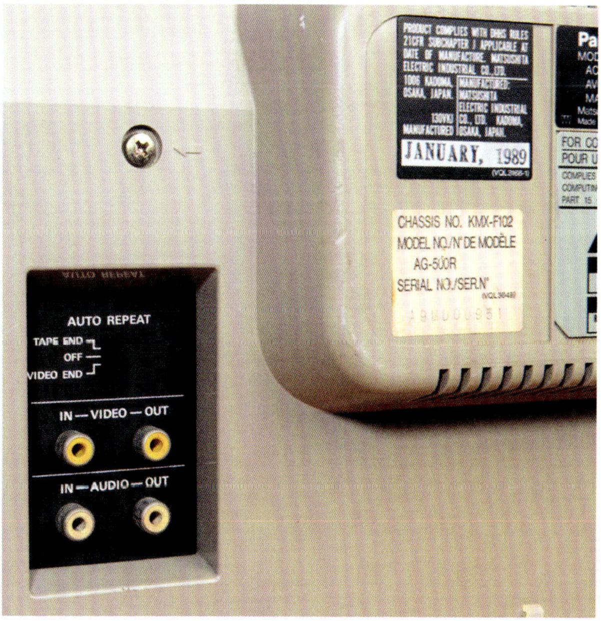
PRODUCT COMPLIES WITH DHHS RULES 21CFR SUBCHAPTER J APPLICABLE AT DATE OF MANUFACTURE. MATSUSHITA ELECTRIC INDUSTRIAL CO.,LTD.
OSAKA, JAPAN.
MANUFACTURED: MATSUSHITA ELECTRIC INDUSTRIAL CO., LTD. KADOMA, OSAKA, JAPAN.
MANUFACTURED
JANUARY, 1989
CHASSIS NO. KMX-F102
MODEL NO./N° DE MODÈLE
AG-500R
SERIAL NO./SER.N°
AUTO REPEAT
TAPE END
OFF
VIDEO END
IN — VIDEO — OUT
IN — AUDIO — OUT

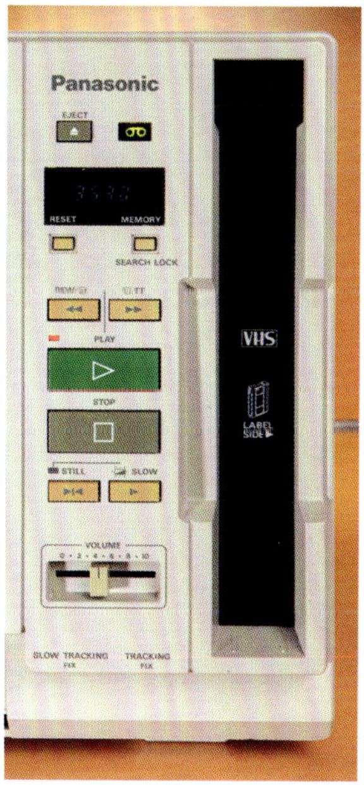
Panasonic
EJECT
RESET
MEMORY
SEARCH LOCK
PLAY
STOP
STILL
SLOW
VOLUME
SLOW TRACKING
TRACKING
VHS
LABEL SIDE

02

THE EARLY DAYS OF THE HOME THEATER

Before everyone had large, flat-screen televisions in their homes, a person was considered rich if they had a "big-screen" projection TV. Whether it be front-projection or rear-projection, seeing one of these was truly captivating. Despite it having a dimmer and fuzzier picture than a direct-view TV, the sheer scale made it feel special. It was the closest thing to being at a movie theater without having to drive across town, spend a small fortune on snacks, and deal with an auditorium full of people who lack self-awareness.

Projection TVs could produce such large pictures because they used lenses and mirrors to enlarge the image. They often had one or more CRTs that had been specially designed to produce an image that would be sent through lenses and reflected by mirrors onto a screen. There are many ways to achieve this, and we will look at a few examples in this section.

Now, you may be wondering why a projection television would require more than one CRT, while a direct-view TV only consists of one. Well, in order to produce a brighter image, the projection system could use a different CRT for each of the primary colors (red, green, and blue). Stacking these three colors meant that a brighter and clearer full-color image could be produced. Therefore, most of the CRT projectors comprised three mono-color picture tubes. The three-tube versions were capable of producing a brighter image not only because three CRTs can produce more light, but also because they do not require a shadow mask/aperture grille like the single-tube models. You will recall from our discussion earlier that the shadow mask/aperture grille prevents a portion of the light from escaping the tube (see page 13).

Different designs include one, two, or three lenses. This refers to the number of visible lenses when looking at the projector. A one-lens system would typically only have one CRT, while a two- or three-lens system would typically have three CRTs to produce the image described above. The two-lens system worked by combining the image from two of the three CRTs using one of the lenses, while the remaining CRT had its own lens. This allowed for a more compact design.

In a front-projection system, the projected image approached the screen from the front. The projector may have been built into the unit with a mirror to reflect the image back to the screen, or it could have been across the room from the screen. For rear-projection units, the projector was often built into the TV cabinet, and the image approached the screen from the rear. The screen would have essentially been a flat piece of translucent glass or plastic panel spanning the front of the cabinet. Rear-projection systems also had the advantage of being more compact.

In the early days of home-theater projection systems, the brightness was often measured in foot-lamberts. This was because fixed-distance projectors were designed with a specific screen in mind. A foot-lambert is a measurement used to describe the amount of light that reaches the viewer from the screen. The lower the value, the darker the room should be for an acceptable picture. Later, the brightness was more commonly measured in lumens, which is the amount of the visible light emitted by the projector. The higher the lumen value, the brighter the image produced by the projector.

In this section, we will look at several types of projection TVs. Although these sets do not usually get much love from CRT enthusiasts due to their enormous size and poorer image quality, there are some models that should be preserved and appreciated. I have accumulated some of these unique and interesting examples to display in the basement.

1975 ADVENT VIDEOBEAM 1000A

Type: 2-piece front projection **Screen size:** 84in **Brightness:** More than 20 foot-lamberts
Weight: Projector: 140lb (63.5kg) Screen with legs: 68lb (30.8kg) **Introductory price:** $3,995 USD

The home-theater revolution began in 1973, when Sony and Advent released the first color CRT projectors to the consumer market. Sony narrowly beat Advent to the market with the VPP-2100, which was a single CRT projector with a 50-in screen, but Advent had the more successful unit, which consisted of three CRTs and an 84-in screen. At the time, a typical color TV screen would have been 19in or less, and the largest direct-view TV was 25in. The Advent VideoBeam 1000 projected onto a massive 7-ft screen, which is more than ten times that size.

Although Advent's goal was to design a low-maintenance, easy-to-operate projection TV, this system would not have been found in your average household. Not only was the original $2,500 price tag a limiting factor for the domestic market, but the amount of space needed for proper viewing also prevented the everyday consumer from purchasing it. Since the projector had a fixed focal distance, it had to be 8ft from the screen, and even the slightest movement of the projector or screen would require it to be re-aligned. Even at such a high price point, Advent still lost money on every unit sold due to the cost of development.

A couple of years later, Advent improved some of the circuitry in the projector and rebadged it as the 1000A. Since Advent had nearly gone out of business, the price was gradually raised to $3,995. There was also a remote-control option for a whopping $4,495! This option had to be requested at the time of purchase and could not be added on later. The remote was basically the full control panel found on the projector built into a box and attached to the projector using a 30ft cable. It sounds absolutely ridiculous, and I would love to come across one. These prices were not exactly consumer-friendly, which resulted in the 1000A primarily still being used in commercial and industrial settings. This technology was, however, owned by Elvis Presley, Hugh Hefner, and other well-to-do clientele.

Let's talk about how the system works. The image is produced by three CRTs, where each tube, called a LightGuide, has a single electron

WHY I LOVE IT

The Advent VideoBeam started the consumer fascination with projection TV, and captured the imagination of those fortunate enough to view one. Elvis had one in his "Jungle Room" at Graceland, Hugh Hefner watched one at the Playboy Mansion, and many other home-theater enthusiasts had rooms dedicated to this groundbreaking system. Although it would be decades before big-screen TVs were commonplace in homes, the first steps were taken by Advent with the VideoBeam, and the home-theater revolution was about to take off.

ADVENT

gun used to produce one of the primary colors (red, green, and blue). Within each LightGuide, a primary color is projected onto a phosphor-coated target then reflected onto a round mirrored surface before being magnified and projected onto the screen. As these three colors are projected onto the screen, a full-color picture is formed. The screen is curved and aluminized to reflect as much projector light back at the viewer as possible while eliminating any excess ambient light. The curved screen also prevented "hotspots," which occur when some parts of the projected image are brighter than others. Since the screen must be a fixed distance from the projector, extensive setup and convergence is required. All the controls are accessible by the viewer, allowing for convergence and alignment as needed. Once setup, it is best to ensure the projector and screen do not move.

The audio for the VideoBeam comes from a speaker mounted below the CRTs and facing away from the viewers. This way the sound bounces off the curved screen, similar to the way the video is reflected. There are also inputs and outputs for both external video and audio sources. Additional speakers could be connected for stereo sound.

"The Advent VideoBeam 1000 projected onto a screen more than ten times the size of the largest direct-view TV available at the time."

1976 SONY TRINITRON KP-4000

Type: All-in-one front projection **Screen size:** 40in **Brightness:** More than 10 foot-lamberts
Weight: 301lb (136.5kg) **Introductory price:** $2,500 USD

By 1976, projection televisions had been around for several years, but they tended to be two separate pieces, with the projector sitting across the room from the screen it projected onto. Sony thought this was impractical for most households because any movement of the screen or projector would distort the image and require it to be recalibrated. Plus, a large amount of space was needed to accommodate these two-piece systems. As it continued to trailblaze, Sony designed an all-in-one projection video system.

That system was the Trinitron KP-4000. It used a 13-in Trinitron CRT specially designed with a lens attachment to produce an image that would reflect off a mirror and project onto a 40-in screen. The entire system was built into a wooden frame wrapped in walnut veneer. The screen and projector are hidden behind a tambour door that can be raised out of sight. With the door raised, the highly reflective, silver screen is revealed, and the projector can fold out to engage viewing mode. To allow the viewer to remain comfortable during viewing, an RM-204 Remote Commander was included, which was a four-button, ultrasonic, wireless remote control. When viewing is done and the door is lowered, the system looks like a beautiful piece of furniture that discreetly conceals the TV. The only hint a television is lurking inside is the iconic Sony Trinitron badge on the control-panel door.

A 40-in screen may not sound like much these days, but, at the time, the typical screen size in a home set was 19in when measured diagonally. This made the 40-in screen more than four-times the size of a typical television set. Although other projection TVs had screens up to a massive 84in, designing an all-in-one system to reflect the image off a mirror and onto a screen significantly reduced the footprint of the system and made setup and use a breeze. The main drawback of these systems was the loss of color and clarity from the CRT to the screen. The coating on the screen was designed to reflect as much of the image as possible while maintaining its brightness, but in order for this to work properly, the ambient lighting needed to be very dim. Otherwise, the screen would also reflect it, causing the image to become washed out and dull. But, hey, we can flip the lights off and have an awesome movie night!

WHY I LOVE IT

I have a soft spot for projection TVs. Even if the image they produce is not as vivid as a regular television set, they still have merit and often come in unique designs. In my opinion, the Sony KP-4000 is the best-looking of the projection TV systems. Not only does it transform to hide the fact that it is even a television, but the walnut veneer and chrome tuning knobs give it all the style cues of the early Trinitrons. The cabinet alone is a piece of art that deserves to be preserved.

SONY

ITEMS OF INTEREST 04

1 **1977 ZENITH PANORAMA VI**

This Zenith (see page 24) was designed with a curved console for corner placement, and it looks really good in the corner, especially when joining two walls of additional TVs together.

2 **1982 RCA COLORTRAK 2000**

This unusual RCA (see page 36) has a sheet of contrast glass across the entire front of the console. When the television is turned off, the set resembles a modern flat-screen TV.

3 **1976 SONY TRINITRON KV-1512**

I saved this 15-in Trinitron from being scrapped by an electronics recycling company. It's a nearly pristine example of this model.

4 **1981 SONY VIDEOSCOPE KP-7220**

While it may just look like a bizarre coffee table, this is actually a three-CRT projector intended to be paired with a 72-in screen (see page 58). Hiding the projector in a coffee table allowed the controls to be easily accessible from the couch.

ZENITH
THE QUALITY GOES IN BEFORE THE NAME GOES ON.
WELCOME TO THE VINTAGE VIDEO BASEMENT
PEPSI
· VHS · BETAMAX · LASERDISC · CED
NOW SHOWIN
MY BLOODY VALENTINE & PHANTASM
THE GOONIES
TURTLES
STRETCH
E.T.
JAWS
CHRISTINE
A Nightmare
TALKING FREDDY
BLOCKBUSTER VIDEO
1
2
3
4

1981 SONY VIDEOSCOPE KP-7220

Type: 2-piece front projection **Screen size:** 72in **Brightness:** 50 foot-lamberts
Weight: Projector: 111lb 9oz (50.7kg) Screen: 37lb 13oz (16.8kg) **Introductory price:** $3,100 USD

The KP-7220 was originally released in 1980, and was Sony's version of a two-piece projection system with a projector designed to be used as a coffee table. That's right—the projector was disguised as a coffee table so it could blend into the environment and serve multiple functions. With pecan wood veneer and a smoked-glass top, it's a stunning piece of electronics décor.

The coffee table houses three mono-color picture tubes—red, green, and blue. These project an image onto a 72-in screen which was placed 8ft across the room. A screen of that size really makes your home feel like a theater. The tubes are "coolant sealed" to prevent overheating from the intense light being projected. The coffee table also allowed the controls to be at your fingertips, without the need for a remote control, while you lounged in your favorite recliner. Not only could you easily change channels and alter the volume, but image adjustments such as brightness, sharpness, and color were also conveniently accessible.

A smaller version was released at the same time. The KP-5020 came with a 50-in screen and required a throw distance of just 6ft, allowing customers to choose the size that best fit their needs and living space. Together they marked the introduction of Sony's Videoscope line that contained most of the projection systems it released throughout the Eighties and Nineties.

WHY I LOVE IT

I mean, it's a coffee table AND a projector! It truly is a perfect piece of décor for a basement where there are already so many electronics. A plain coffee table wouldn't be nearly as cool. Also, it really is a beautiful design that fully captures the time period in which it was released.

OUIJA
SONY

1984 SONY VIDIMAGIC FP-60

Type: Portable Indextron tube front projection **Screen size:** 30–200in **Brightness:** 50 foot-lamberts on a 55-in screen **Weight:** 34lb 3oz (15.5kg) **Introductory price:** $2,995 USD

The Vidimagic was among the Sony projection systems that was not designated as part of the Videoscope line during the Eighties. Now, I'm cheating a bit by including this projector in the home-theater section of the book, because the Vidimagic was intended to be industrial/educational equipment. I'm going to justify it by pointing out that this was an early example of projectors becoming small and portable. The projection systems we have explored to this point have been fixed-distance systems that were intended to be set up and left undisturbed.

The Vidimagic is special because it has a single 5.25-in high-brightness Indextron picture tube that enables it to be refocused, so picture size and screen distance can vary based on the needs of the viewer. The projection size ranges from 30in to a massive 200in! As the screen size increased, the throw distance increased, and, therefore, the image brightness decreased. No special, curved screen was required either. Any flat, white surface such as a wall could be used as a screen. For easy positioning, the adjustable, built-in stand allowed the angle of the projector to vary between 7 and 20 degrees.

The FP-60 has a built-in Betamax VCR for BII and BIII use. It is surprising Sony did not include BI capability, since this unit was intended for professional use, but the inclusion of the Betamax is still impressive. There was also the FP-62 model that did not include the Betamax. Both versions had a built-in tuner for watching TV. A remote control/microphone that allowed the projector to be used as a public address (PA) system was available as an optional accessory.

An interesting note about the Vidimagic is that the cosmetics company Avon purchased many of the FP-60s for its door-to-door salespeople to use. They would set up the Vidimagic in their clients' homes and play a tape showcasing Avon products. These machines were even branded with the Avon logo and were supplied with a carrying case, speakers, and portable screen. I am still on the lookout for one of these sets in good condition.

WHY I LOVE IT

A portable CRT projector with built-in Betamax from the mid-Eighties is simply a technological marvel. Its design gives it the look of a video canon ready to blast movies onto the big screen. It is still fun to take the Vidimagic outside for movie nights. Sure, there are many modern projectors offering better and brighter images, but it is hard to beat the aesthetic of a Vidimagic!

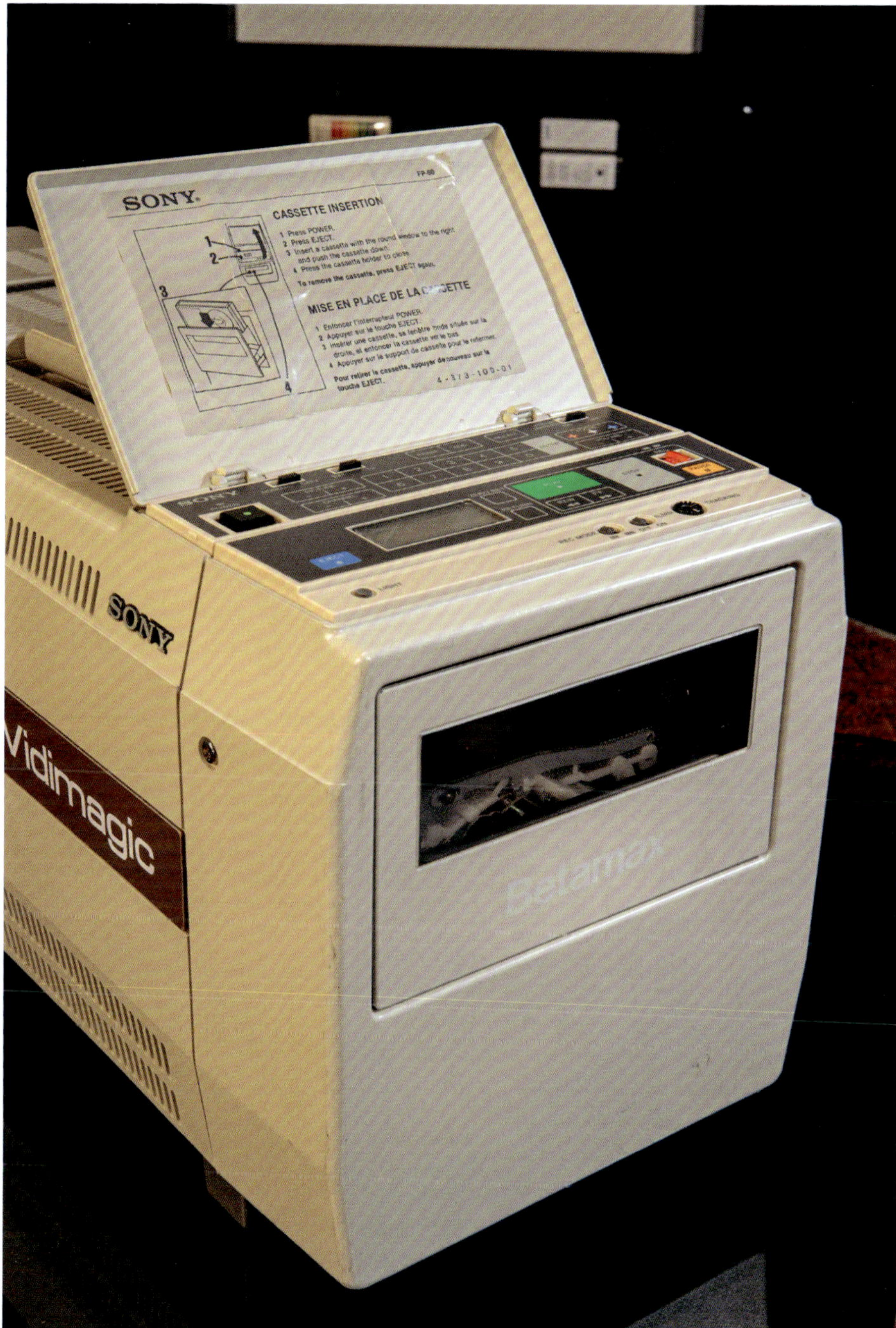

1984 SONY VIDIMAGIC FP-60

1997 SONY VPH-D50

Type: Front or rear projection **Screen size:** 60–250in **Brightness:** 800 lumens
Weight: 119lb (53.9kg) **Introductory price:** $9,995 USD

Let's jump forward over a decade, well into the Nineties, to see how the CRT projector had evolved. In 1997, Sony released the VPH-D50 as the entry-level model, along with the higher end G70 and G90 models. By this point, CRT projectors were capable of displaying HDTV, and the VPH-D50 had a maximum resolution of 1280x1040p. CRT projectors were still typically composed of three CRTs which, on this model, were 7-in tubes.

Although this VPH-D50 weighs close to 120lb, it was designed with mobility in mind. Handles pop out on both sides of the projector to allow two people to maneuver it easily, and the handles then conveniently slide and latch back into place. The projector could also be set up in a variety of positions and heights—from being placed on a low table to mounted upside down on the ceiling. You could even put it behind the screen and use it for rear projection. Technology had come a long way from the days of fixed-distance systems. The VPH-D50 was capable of projecting an image from 60in to a whopping 250in! By this time, special screens, such as those required by the early CRT projectors (see pages 50–61), were no longer needed, so the freedom and options afforded by the new technology greatly improved the home-theater experience.

WHY I LOVE IT

This projector produces a beautiful image, and, while there are better machines, the VPH-D50 showcases the improvements being made with CRT technology before CRTs faded from production. According to the person I bought this unit from, the projector was found in a barn that his company was dismantling. After years of it being in the way, the guy was ready to get rid of the cumbersome unit. He sold it to me for $50. It now gets regular attention and provides big-screen enjoyment.

ITEMS OF INTEREST 05

1

1975 SONY TRINITRON LV-1901

Sony introduced the Betamax as part of a console with a 19-in Trinitron. This complex-looking system (see page 72) was targeted at the ordinary consumer and proved to Sony there was a market for VCRs in the home.

2

SPEAKERS

Two speakers are hidden below the picture tube. One is a 6¼-in cone-type speaker, and the other a 2¾-in tweeter.

3

SONY BETAMAX SL-6200

The very first Betamax released in the US was easily accessible from the top of the LV-1901 (see page 72) and could be protected by the hinged dustcover.

4

STORAGE

A secret compartment next to the speakers can hold all of your Betamax tapes and help you organize your recordings.

3
1
SONY
9 43
2
4

ZENITH
THE QUALITY GOES IN BEFORE THE NAME GOES ON.
VINTAGE VIDEO BASEMENT
THE GOONIES
I'm here for the VCRs
THE KARATE KID
GHOSTBUSTERS
PIECES
SONY
VIDEO CASSETTE
K-60
STAR WARS
RETURN of the JEDI
THEY LIVE
SCREAM
Happy Gilmore
SONY Betamax

03

JOURNEY THROUGH THE FORMAT WARS

Now that we have explored many styles of television, let's delve into how the use of the television evolved with the introduction of the VCR. Traditionally, TVs were solely used for watching broadcast television. A handful of channels were picked up by an antenna and displayed on the screen. However, the Seventies and Eighties expanded TV's capabilities, with the introduction of new accessories such as video games, video cameras, and, of course, the VCR.

Videotape was invented in 1951, and Ampex released the first commercially available videotape recorder (VTR) in 1956. Early VTRs were reel-to-reel style: a reel containing the video had to be mounted onto the machine and hand-loaded along its path. An empty reel captured the videotape as it passed through the recorder. Incorrect loading and excessive handling of the tape could easily result in damage. Therefore, a better method for protecting the tape was needed, and the videocassette was introduced.

VCR stands for "video cassette recorder," and refers to any machine that records on a videocassette, whether it's VHS, Beta, or one of the lesser-known formats. Therefore, all VCRs are VTRs, but not all VTRs are VCRs. Eventually, "VCR" became the dominant descriptor for machines that utilized cassettes, and it is the one we will stick with to reduce confusion.

As a huge movie fan, I absolutely adore VCRs because they introduced a method for watching your favorite movies at home on your own time. That may sound like a primitive idea today, when you can stream content straight to the phone in your pocket, wherever you are and whenever you want, but it was revolutionary at the time. Nowadays, it's possible to stream almost anything, anywhere, at any time. But, for decades, being able to pop a new release into your VCR was a highlight of many Friday nights. There is something magical about scanning a wall of physical videocassettes, selecting a movie based on the cover art, and then holding it in your hand. It is so much more meaningful than scrolling through thumbnail images and mindlessly pressing "play."

In this section, we are going to explore these magical machines for physical media. More specifically, we are going to journey through the golden era when VCRs were being released, and investigate the war for home-theater dominance that was fought between the formats. Instead of giving the same summary about the format wars that is regurgitated endlessly—and is often wrong about the details—we are going to look chronologically at many of the VCRs in my collection to better understand the real story of how the format wars unfolded.

1980 SONY U-MATIC VO-2610

Format: U-Matic (¾in) **Weight:** 66lb 2oz (30kg) **Introductory price:** $2,150 USD

In 1971, before VHS and Beta, Sony introduced the first VCR to the industrial and educational markets. This professional format used ¾-in tape and was, therefore, often simply referred to as "three-quarter inch." Sony branded its line "U-Matic," but the format was also produced by several other companies including JVC (U-VCR) and Panasonic (U-Vision). The "U" refers to the U-shaped path the tape follows during recording. This single-speed system only allowed for one hour of content per cassette.

We will talk more about speeds as we get into the format war between VHS and Beta, but a single speed means it was not possible to increase the amount of video that fitted onto a cassette; an hour of recording was the maximum that was possible. However, the ¾-in tape initially allowed for recording quality that was approaching broadcast level at 240 lines of horizontal resolution, and it could capture stereo audio. In the early days of U-Matic, an hour might have been the maximum recording possible per tape, but tapes that held just 30, 20, or even as low as 10 minutes of content were also available. Prices ranged from $17 for the ten-minute tapes, to $35 for the one-hour tapes. I think it is important to discuss this to better understand the frame of mind Sony was in when it released the Betamax with just an hour of possible record-time. It will turn out to be poor foresight of the different ways consumers and professionals would want to use VCRs.

Sony intended the U-Matic to be a consumer unit, which can be seen with the original U-Matic VCR, the VO-1600. This had a built-in VHF/UHF TV tuner to allow for home recordings. Although there were some videophiles who purchased a U-Matic VCR for home use, the price of the machines was out of reach for the typical consumer. Of course, no prerecorded movies were available to purchase either, so all-in-all, the U-Matic was not fit for home use. Later models dropped the built-in tuner/timer, and Sony focused on the professional market. The format succeeded for decades in the professional environment, and I even have some electronic press kits on U-Matic for movies that were released in the early 2000s!

The player in my collection is the VO-2610 model from 1980, and it has the external tuner/timer that was an optional purchase. The top of the player is adorned with colorful buttons, knobs, switches, levers, and even VU meters that make this U-Matic a joy to operate. When compared to consumer-level VCRs, it is easy to distinguish as a professional machine. It is a top-load device and looks similar to other U-Matics produced by Sony during the first decade of production, although front-load versions did become available in the early Eighties.

"Although there were some videophiles who purchased a U-Matic VCR for home use, the price of the machines was out of reach for the typical consumer."

WHY I LOVE IT

The U-Matic is special because it is the original videocassette format. Although the machine I have is not one of those first models, it still has that classic design. I got it from a foundation in Bellevue, Washington, that used it to make promotional videos to recruit new members. I even have one of the original videos from the Eighties, which shows the VCR in the background of an office scene. Being able to see a recording of my U-Matic in its prime on a tape it edited is something special.

U-matic
PUSH
SONY
MEMORY COUNTER
AUDIO MONITOR
PROGRAMMED OPERATION
REPEAT
POWER
CH-1
START
OFF
MIX
PUSH ON
CH-2
REWIND
HEADPHONES
SONY
VIDEOCASSETTE RECORDER

CAUTION
ATTENTION
REMOTE
VIDEO IN
CH-2/R
CH-1/L
LINE IN
IN
TV
RF OUT
AUDIO MONITOR
COLOR LOCK
AC IN
VIDEO OUT
CH-2/R
CH-1/L
LINE OUT
OUT
SONY
VIDEOCASSETTE RECORDER

1975 SONY TRINITRON LV-1901

Format: Beta **Weight:** 288lb 2oz (130.7kg) **Introductory price:** $2,295 USD

While *Jaws* was still in theaters, becoming the first blockbuster of all time and changing the way the world thought about movies, the introduction of the consumer VCR was also about to cause a revolution. It was not a new concept at this point. The U-Matic had been around for several years in professional and educational environments, but a market for consumer-level VCRs was emerging.

The first successful VCR on the scene was the Betamax by Sony. Taking much of what it had already done with the U-Matic, Sony made the Betamax more compact by using ½-in videotape. Initially, Sony's sole offering was the Trinitron LV-1901. This all-in-one system included a 19-in Trinitron, which was the largest available at the time, accompanied by an SL-6200 Betamax. The Betamax was concealed in the console and a hinged dust-cover lid allowed access to the player. The system required the customer to either replace their current TV or buy an extra set. A stand-alone Betamax became available the following year.

The SL-6200 Betamax was unusual because it did not have a built-in tuner or clock. That portion of the system was built into the console of the LV-1901 with the other controls. As a matter of fact, the console contained two tuners. You have to keep in mind that, at this point, the primary function of the VCR was to allow the consumer to watch television programs at their convenience and no longer be controlled by the programming schedule. It was referred to as a "time-shifting" device, and gave the user the power to record a show while out of the house, or even record one show while they were watching another.

Before this machine entered the home, if you missed an episode of your favorite show, you would either have to wait for reruns during the summer, or possibly never get to see it. It is difficult to imagine this struggle in the modern age of streaming, but the desire to watch what you wanted when you wanted was in the process of becoming a reality.

"Enough units of the LV-1901 were sold to convince Sony there was a market for consumer-level VCRs, making this the first shot in the war for home-video dominance."

SONY
9 43

At the time, broadcast TV in North America produced a picture with a horizontal resolution of 525 lines, and the Betamax had a resolution of 250 lines. The original Beta tapes Sony released were the K-30 and the K-60. The number indicates how many minutes of playback were allowed by each tape. In plain terms, the K-30 was intended to record a 30-minute program, while the K-60 was suited to a full hour episode. This had the major drawback that multiple tapes were required to record a full movie or sporting event. It seems as though Sony, coming off the success of the U-Matic in the professional setting, assumed an hour of recording would be adequate for consumers as well. Unfortunately, this would not be the case. The tapes were not cheap—one K-30 retailed for $12.95 and the K-60 had a price tag of $16.95—and one of Sony's biggest missteps had occurred.

Additional features of the LV-1901 include a flip clock, speakers concealed in the base, video camera input, and a hidden compartment to store your Beta tape recordings. Although the LV-1901 was too expensive and, perhaps, too impractical for the ordinary household, enough units were sold to convince Sony there was a market for consumer-level VCRs, making this the first shot in the war for home-video dominance.

WHY I LOVE IT

The LV-1901 is a jewel in my collection. It contains two of my favorite things—a Trinitron and a Betamax. The fact that it has the very first Betamax, which was the first successful VCR for the home, makes it all the better. It is a beautifully designed machine and an indication that Sony saw a future where the VCR would be just as prevalent in the home as the television. I found this one in an antique store in Oregon. Although it had been for sale at one point, the owner had decided he would rather keep it for display. After several hours of negotiations, the historic piece came home with me.

SONY
authorized service station
SONY
VIDEO CASSETTE
K-60

1976 SONY BETAMAX SL-7200

Format: Beta **Weight:** 45lb 8oz (20.7kg) **Introductory price:** $1,295 USD

When Sony introduced the first standalone Betamax, the SL-7200, it looked similar in style to the SL-6200. The biggest change was the built-in tuner. This addition added turret-type tuning knobs to the front of the machine, which, of course, allowed the consumer to tune the VCR to the channel they wanted to record. Sony did not, however, include a timer, which is surprising because it is difficult to "time-shift" without one. Now the reasoning behind this was that Sony was using flip clocks at the time and did not want a fault with the clock to require the entire Betamax to go in for service. Because of this, it made the timer a separate accessory. Programs could still be recorded without the timer as long as the viewer was present to start the recording, but often times the reason for recording a program was because the viewer would be unavailable for its airing. With the external timer, only one unattended program could be recorded over a 24-hour period. The fact that Sony did not include a built-in timer was another misstep that JVC would soon avoid with its VHS VCRs.

The Betamax loaded a tape using a U-load method similar to that used in the U-Matic machines. During the loading process, 24in of tape are removed from the cassette. This may seem like an excessive amount, but an extended length of tape allowed for less tension and therefore less chance of stress damage to the tape as it moved through the machine. A highlight of the system is that it allowed the tape to remain threaded during actions such as rewind and fast-forward, reducing the need for the tape to be constantly threaded and unthreaded.

Betamax
SONY
PROGRAM SELECT
CASSETTE
EJECT
REWIND
STOP
PLAY
F FWD
RECORD
AFT
VHF
UHF

Just like the LV-1901, the SL-7200 was also only capable of a maximum of one hour of recording per tape. Sony valued the compactness of the Beta cassette over the desire of the consumer to record longer programs such as movies or sporting events, or to record multiple programs on the same tape. In fact, Sony anticipated that consumers would own just a handful of tapes and, after watching a recording on one, would use it to record another program. At this point, little thought had been given to the idea that people would store a collection of recordings to watch over and over, or that there would soon be a market for Hollywood to sell its movies on videocassette. Although the recording quality on Beta was similar to that at the professional level, the size of the cassette would never allow a full-length movie to fit on a single tape at this quality. Even so, the Betamax was a success, and Sony sold more than 100,000 units during the first year of sales. The home-video revolution had begun!

WHY I LOVE IT

This is absolutely my favorite VCR style of any format. In my opinion, the aesthetic styling of the VCR only went downhill from here. Not that other VCRs do not look great—this is just the one that most appeals to me. The wood-grain sides, big rotary knobs, and piano-type transport keys make it a mechanical marvel. The tactile nature of all these knobs, switches, and keys is why I love these vintage video machines. Modern-day touchscreens do not hold a candle to how enjoyable it is to operate this VCR. I happened to come across one of these in an antique store while on vacation in Hawaii. It was such a surprising find, and I had to take advantage of the situation to get some incredible photos at the beach. It was all fun and games until I had to figure out how to smuggle this 45-lb beast back to the mainland. Let's just say the TSA was not excited about this mechanical monster being loaded into the cargo bay.

Betamax
EJECT
REWIND
STOP
PLAY
F FWD
RECORD
PROGRAM SELECT
CASSETTE
COUNTER
MEMORY
ON
OFF
SONY

SL
SONY
SL-7200 3CH
SONY

ITEMS OF INTEREST 06

HOLLYWOOD VS BETAMAX

Just as the format war was getting underway, Sony and Betamax were also going into battle in the courtroom. In 1976, Walt Disney Productions and Universal Studios sued the Sony Corporation over fears Betamax would cut into their profits. They felt the Betamax would infringe their copyrighted programs. The case first went to court in Los Angeles in 1979, and the verdict was in favor of Sony and the rights of consumers to record programming from their TVs. Recording free broadcasts for later viewing in the home was deemed "fair use" of the programs. However, in 1981, the San Francisco Appeals Court overruled this decision on the basis that home recordings infringed the copyright of the producers, and Sony was guilty of "contributory infringement." Congress had passed the Copyright Law that stated only a copyright owner could make a copy of their property. Videotaping is the act of making a copy, so every time a consumer recorded a program, they were breaking this law. In fact, many VCR print ads at the time included fine print stating that recording copyrighted programs may infringe on the rights of the copyright owner. They were essentially telling prospective customers that using the VCR as intended might be illegal!

At this point, the entertainment industry was not as concerned with home recordings as it was with getting a share of all sales related to home recordings. It didn't want to ban VCRs or fine the consumer, it wanted royalties. The studios not only wanted consumers to purchase prerecorded tapes from them, but they also thought they should get a percentage of every blank tape and VCR that was sold. Sony disagreed and appealed the decision. It was clear the case would have to be pushed all the way up to the US Supreme Court.

During this time, VCRs and blank tapes were selling rapidly. Even if illegal, home recordings were still happening with little consideration of the law. After an eight-year battle and a nail-biting five-four decision, the Supreme Court announced its ruling to legalize home video recording in January 1984. As long as the consumer was not trying to profit from their recordings, they had the right to videotape what was being broadcast. Although Betamax had won this battle—one that would benefit other home video formats—its biggest fight, against VHS, was still underway.

Beta is Better
SONY
BETA IS BETTER
ZENITH

1977 JVC VIDSTAR HR-3300U/HR-3600AU

Format: VHS **Weight:** 29lb 11oz (13.2kg) **Introductory price:** $1,200 USD

Although the VHS system was not the first VCR to hit the market, it had the advantage of learning from previous systems, and it made improvements that would lead it to dominate the home-video market. One of those improvements was the increase in cassette size. A VHS is approximately 30 percent larger than a Beta cassette. Although it would seem as though a smaller cassette would be a plus, the fact that more tape could fit into a larger cassette, and therefore more content could be recorded, turned out to be a big plus for consumers, and a huge advantage for VHS.

The first VHS VCR to hit the international market was the HR-3300, launched under the Victor badge in Japan in 1976. Practically the same machine was released in the US the following year as the JVC HR-3300U. The JVC Vidstar had a two-hour recording capability. Right out of the gate, VHS was capable of recording for twice as long on a single tape as Beta. This allowed an entire movie to fit on a VHS, and would be the single most important reason why VHS won the format war against Beta.

VHS systems used the "M-load" method for loading the tape from the cassette. The basis for this method was patented by Sony, but since it opted for the U-load technique for Betamax, the organization licensed it to JVC and its parent company Matsushita Electric. JVC further developed it to fit its VHS systems. This required just 13in of tape to be removed from the cassette during loading, but subjects the tape to more stress. It also required constant loading and unloading during mode changes, such as going from rewind to play, which made the system more time-consuming.

REC
REW
STOP
PLAY
FF
EJECT
COUNTER
SEARCH
ON OFF
UHF
VHF
AFT
ON OFF
POWER
REC SELECT
TV
AUX
VIDEO/TV
VIDEO
TV
TIMER/SLEEP
ON
OFF
MIN
HR-3600AU SPEED PLAY VHS

The JVC Vidstar also had a digital timer built into the lower-left corner of the console, which was more convenient and streamlined than the external flip clocks that were being used by Sony. This was yet another easy-to-see improvement over Betamax. Even though Betamax allowed for higher-quality recordings, it is much easier for a consumer to see the difference between playback lengths than it is to compare the quality of recordings. Quality does not solely depend on the abilities of the VCR, it also depends on external factors such as the input signal of the broadcast and the age and type of TV the program is being viewed on. This makes it more of a subjective variable between the formats and, it turns out, besides videophiles, most people did not care or could not tell the difference between VHS and Beta.

Along with the HR-3300U, I also own an HR-3600AU, which includes several features not found on the 3300U. This version began the war of "special effects" by giving you the ability to control the image speed. The JVC HR-3600AU introduced both freeze-frame capabilities, which allow you to pause a video and still be able to inspect the picture, and double-speed playback, which allows you to progress through a video at twice the speed. There is a button above the clock display labeled "Speed Play," which enables the machine to play a video at double-speed. Next to it is a port for plugging in a wired remote control that gives you access to this feature from the comfort of your seat.

WHY I LOVE IT

These JVC Vidstars are my favorite of the VHS VCRs, for the same reason I love the Sony Betamax SL-7200.. JVC created a beautiful machine that demanded attention when displayed in your home. The diving board-style control levers are a standout feature on this sleek machine, and even though it was a basic device with minimal features, it looks as though a graduate degree was needed to operate it.

JVC VIDSTAR VHS
PAUSE/STILL
AUDIO DUB
REC
REW
STOP
PLAY
FF
EJECT
COUNTER
UHF
VHF
SPEED PLAY
REMOTE CONN
SEARCH
AFT
SELECT
TIMER SET
TIME ADJ
HOUR
MIN
REC SELECT
TV
AUX
VIDEO/TV
VIDEO
TIMER/SLEEP
POWER
HR-3600AU SPEED PLAY VHS
MIC
JVC VIDSTAR VHS
PAUSE
AUDIO DUB
REC
REW
STOP
PLAY
FF
EJECT
COUNTER
UHF
VHF
SEARCH
AFT
SELECT
2025
HOUR
MIN
REC SELECT
VIDEO/TV
TIMER/SLEEP
POWER
MIC

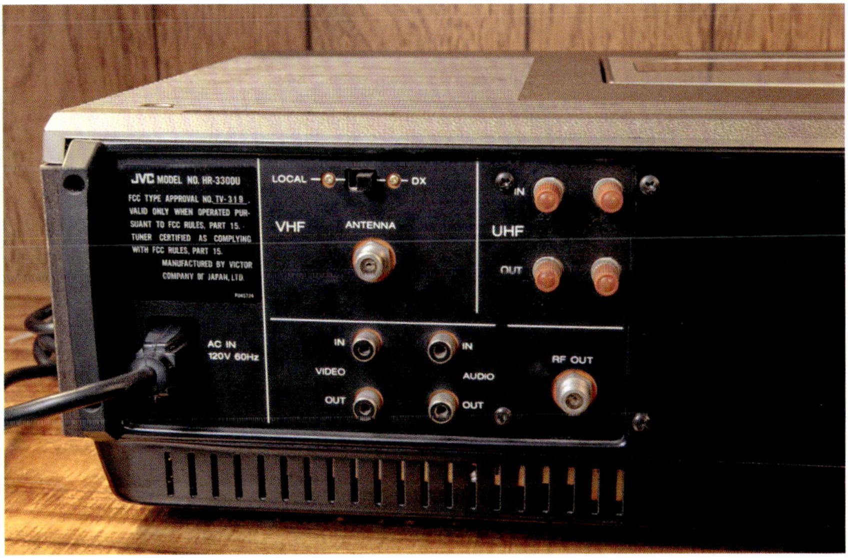
JVC MODEL NO. HR-3300U
FCC TYPE APPROVAL NO. TV-319 . VALID ONLY WHEN OPERATED PURSUANT TO FCC RULES, PART 15. TUNER CERTIFIED AS COMPLYING WITH FCC RULES, PART 15.
MANUFACTURED BY VICTOR COMPANY OF JAPAN, LTD.
LOCAL
DX
VHF
ANTENNA
UHF
IN
OUT
AC IN 120V 60Hz
IN
VIDEO
OUT
IN
AUDIO
OUT
RF OUT

1977 RCA SELECTAVISION VBT-200

Format: VHS **Weight:** 38lb 4oz (17.4kg) **Introductory price:** $1,000 USD

The RCA SelectaVision VBT-200 was the first VHS VCR released in the US. JVC had previously released the HR-3300 under the Victor badge in Japan in 1976, but RCA made it to the market in the US first, and it really came out swinging. The RCA SelectaVision allowed for a whopping four-hour record time when placed in LP (Long Play) mode. In fact, RCA only agreed to back VHS because it was allowed to extend the recording length. This was another big advantage for VHS.

The JVC could supply a maximum of two hours of recording/playback due to only being capable of Standard Play (SP), while Beta still only allowed for just one hour of recording per tape. This was quite an advantage for VHS, which could now record four times more than Beta per tape. Although the longer recording time resulted in diminished quality, it would soon be seen that the common user was more concerned with the high price of tapes than the slightly lower quality image produced by longer record times. All the other VHS brands soon followed suit and offered the longer play feature in their machines, although JVC initially held out in an effort to maintain video quality.

Another great feature introduced by the SelectaVision was a corded pause remote, which made editing out commercials while recording easy and convenient. The Betamax required the user to get up during every commercial break to avoid recording any commercials, unless they had stationed the recorder next to their chair. Additionally, the clock/timer was built in, similar to the one in the JVC Vidstar. These great features caused many prospective VCR customers to look away from Betamax and toward VHS. Overall, VHS had landed several major blows against Sony and the Beta format.

WHY I LOVE IT

Although the RCA SelectaVision did not have the style of the Sony Betamax or the JVC Vidstar, it was considerably cheaper and showed the capability of the VCR and how forward-thinking the companies behind VHS were. Altering the record speed to allow even more content to be recorded per tape became a staple in the format war, and would soon allow Beta to fit a full-length movie onto a single tape.

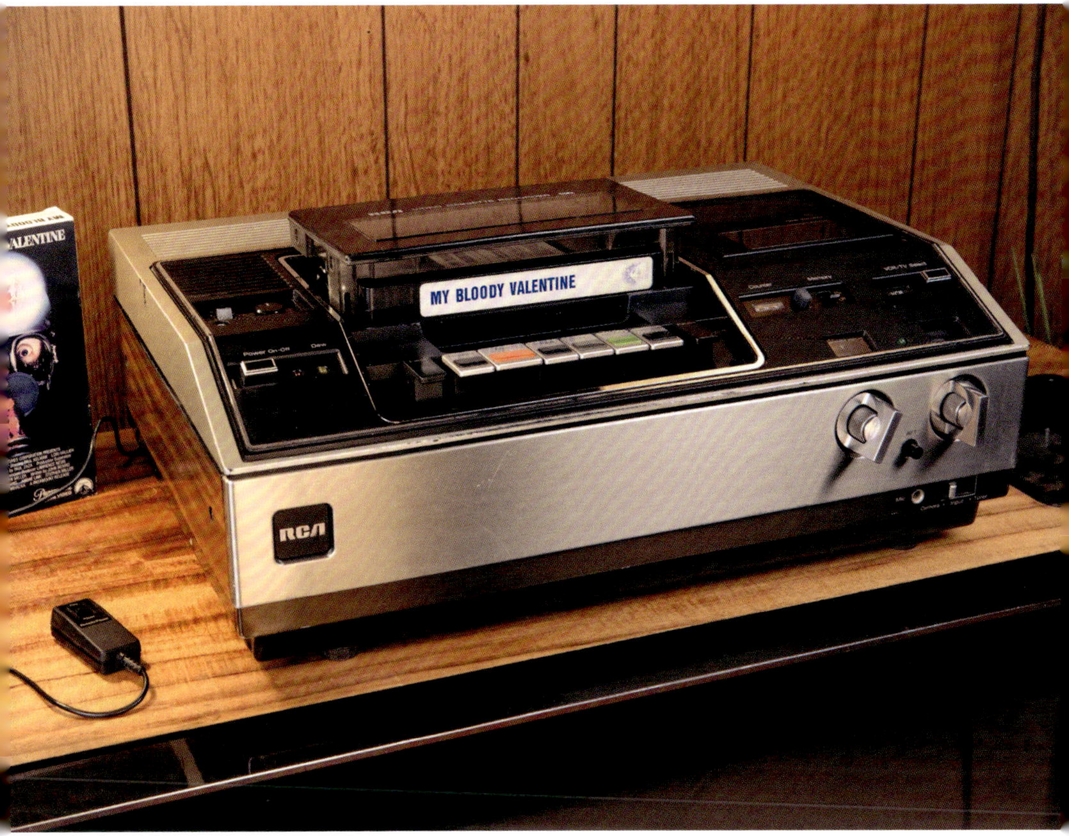
VALENTINE
MY BLOODY VALENTINE
Power On-Off
RCA

1977 QUASAR VR-1000

Format: VX **Weight:** 44lb (19.9kg) **Introductory price:** $995 USD

VHS and Beta were not the only entrants in the format war. Quasar, a Matsushita Electric company, threw its hat into the ring with the VR-1000. As early as 1975, Matsushita had released other VX-format players in other parts of the world, before focusing on the VHS format. Better known as the "Great Time Machine," the VR-1000 was Quasar's attempt at becoming a contender with a different videocassette format. Similar to the VHS system, the Great Time Machine was capable of recording up to two hours per tape, and could record the program you were watching or a different channel to the one you were watching. However, an optional $49.95 timer was needed to record programs while you were away from the TV. That's right: the Great Time Machine didn't even come with a timer, which seems counterintuitive! With the timer, it was the only VCR on the market in 1978 that allowed the user to set up recording for more than one program in a 24-hour period. This made it possible to record multiple programs on the same cassette while the user was away. The programs did, however, have to be shown on the same channel.

Like other VCRs of the time, the Great Time Machine was of the top-load variety, but instead of the cassettes being inserted wide-side first like a VHS, they were inserted narrow-side first. Once in place, a long lever that ran across the front of the machine was pushed to the far right and loaded several mechanisms into position.

VX cassettes consisted of two reels stacked on top of each other, unlike the side-by-side placement in a VHS. When inserted into the VCR, no tape was removed from the cassette. Instead, the video drum entered the reel well from the underside of the cassette.

The cassettes were considerably larger than both the VHS and Beta alternatives. In fact, they rivaled the U-Matic in size. There were two notches in the side of the shell for the user to place their fingers during insertion and removal. VX cassettes retailed for $18.95, $27.95, and $33.95 for the 60, 100, and 120-minute tapes, respectively. Quasar soon ditched the VX format and joined the rest of the Matsushita family in supporting the VHS format. The VR-1000 would be the only VX-format VCR released in North America.

WHY I LOVE IT

I cannot help but be drawn to this unusual format—it's such an odd design and so different from VHS and Beta. The cassettes are large and impractical, the machine is bulky, and the lever used for loading and ejecting the cassettes is one of the craziest things I've seen on an early VCR. It was doomed to fail from the start, but the fact that any format and machine design was possible in those days is immensely interesting to me.

VIDEO 120
CASSETTE
VC120
Quasar
VR-1000

START
SLIDE SLOWLY
VR-1000

ITEMS OF INTEREST 07

ADDITIONAL VICTIMS OF THE FORMAT WARS

The Quasar "Great Time Machine" was not the only videocassette format that threw its hat into the ring and got pummeled. There were many others, and some even predated VHS and Betamax. Here is a list of some of the lesser-known contenders in the format war.

1 CARTRIVISION

The short-lived Cartrivision was, initially, a TV set with a built-in recorder that was sold at just a handful of Chicago-area department stores, such as Sears and Montgomery Ward, before expanding into more than 250 outlets. The price for the console was $1,350. In 1973, after just over a year, the Cartrivision was discontinued due to poor sales and the minor issue that all the videotape had already started to fall apart! An interesting part of Cartrivision history is that feature films were available for rent on red videotape cartridges. These red cartridges could only be watched once before being returned to the retailer, because a special machine was needed to rewind them. The Cartrivision was, however, capable of rewinding the black cartridges that were purchased for recording. A video camera for recording home movies was also available.

2 PHILIPS VCR

Philips released a consumer-level VCR system as early as 1972. Yes, it was simply called "VCR." Philips even trademarked the initials. The cassette consisted of two reels stacked one on top of another, instead of the more common side-by-side layout we are used to with VHS. Although it predated VHS and Betamax, the Philips VCR was only marketed in a few countries, including the UK, Europe, and South Africa. While it did have some success, it never expanded to North America and came to an end after less than a decade.

3

PHILIPS VIDEO 2000

Philips followed up its VCR system with a much-improved format called Video 2000, which was released in 1979. A cool feature of this format is that the cassette it used looked like a large audio cassette and housed ½-in videotape. Only half of the tape width was used during recording, so, similar to an audio cassette, it could be flipped over for recording on the other side as well. The Video 2000 was also never released in North America and, by 1988, had succumbed to the impact of the VHS.

4

SANYO V-CORD

The V-Cord was released in 1974. Even though it was introduced before Betamax, it only recorded in black and white and used the skip-field method similar to the Cartrivision. The skip-field method is a way of recording only half of a frame of video because only one of the two fields that make up a frame is captured. This method results in a lower-quality recording than those captured by the more sophisticated VCRs using Beta and VHS formats. An interesting design choice of the V-Cord is that the cassette was inserted narrow-side first, but the tape was pulled from the long side of the cassette to interact with the video drum. The improved V-Cord II was released in 1976; it recorded in color and captured both fields of the image. Sanyo would soon join Sony in producing Beta format machines, which it named "Betacord."

1978 RCA VCT-400

Format: VHS **Weight:** 34lb (15.4kg) **Introductory price:** $1,275 USD

RCA continued to make advances and introduced the first VCR capable of programmable recording in 1978. Previous VCRs may have been capable of recording multiple programs while unattended, but they all had to be on the same channel. The VCT-400 allowed for up to four programs on different channels to be recorded throughout an entire week. This was an impressive feat. Want to go on vacation for a week and not miss your favorite shows? The VCT-400 had you covered.

It did this by doing away with the giant tuner knobs and incorporating electronic channel tuning and push-button presets. To set up a recording, the user had to set the start time, the day, the length of recording, and the TV channel. It was not an intuitive process, however. For example, to set the day you couldn't merely press a button that corresponded to the day of your choice, because the machine didn't know what day it was. Let's say today is Sunday and you wanted to record something on Wednesday. You had to count up from the current day, starting at zero, so that meant setting the day to "3." If you weren't careful, it was easy to record the wrong show. Even still, the VCR was becoming a more sophisticated appliance, and VHS was beginning to lead the way.

WHY I LOVE IT

It's clear that RCA played a major role in the success of VHS in the United States. It could claim to have the first VHS VCR on the market, the first VHS VCR to have multiple speeds allowing for up to four hours of record time, and the first VCR with programmable recording capabilities.

SUPERTAPE
T-120 2-4-6 HOUR
VIDEOCASSETTE
Radio Shack
RCA
Counter
Memory
Power
Dew
5:35

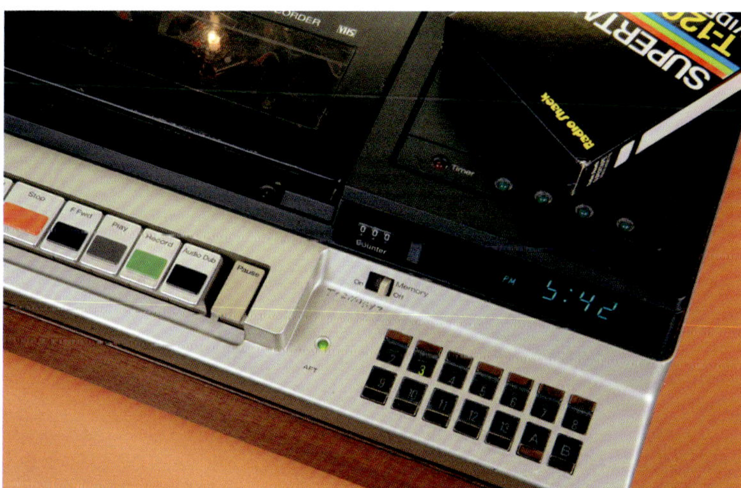
Stop
F.Fwd
Play
Record
Audio Dub
Pause
Counter
Timer
Memory
On
Off
AFT
5:42
Radio Shack
A
B

1978 SONY BETAMAX SL-8200

Format: Beta **Weight:** 45lb (20.4kg) **Introductory price:** $1,260 USD

Sony released the SL-8200 in 1978 to compete with the longer recording capability of the VHS. Even though it looked very similar to the SL-7200, a switch on the top of the console allowed the speed of the tape to double in order to fit twice as much content. Keep in mind that this also causes the quality to be slightly diminished. Betamax now offered up to two hours of record time, but the VHS had already jumped to four hours.

To combat this, Sony introduced an automatic tape changer that attached to the top of the Betamax. During recording, when a tape reached the end, the changer would eject the first tape and insert the second tape. This allowed the Betamax to essentially record for the same length as VHS. There is just one major difference—the Betamax still used two tapes for every one VHS. Although a full movie could now fit on a single Beta tape, it turned out consumers preferred to fit multiple programs onto one tape if it cost them less.

Sony still did not bother to incorporate a built-in timer; it remained a separate accessory. I actually love the way the flip clock looks on top of the SL-8200. Aesthetically, it is a beautiful combination that makes a statement, but I imagine that at the time it might have been inconvenient when compared to the sleek design of a built-in timer found on VHS systems. Additionally, there was also still no remote control for editing out commercials while recording. Some advertisements by Sony tried to sell the idea of placing the VCR next to your recliner for easy access to the keys without getting up.

The SL-8200, along with the Zenith JR9000 and the Toshiba V-5210, was the transition point between X1 and X2 recording speeds—later to be renamed BI and BII. These machines were capable of recording and playing in both speeds, but moving forward, the Betamax consumer would essentially lose BI capabilities as it was deemed to be a professional format. At the time, it was becoming obvious that the consumer cared more about tape economy and prices than quality and cassette size. Since BI was stuck at a maximum of one hour of recording time, it was all but removed from consumer machines. After this point, tapes recorded at the BII speed would not play on a BI machine. Practically all prerecorded tapes of movies or events would be released at the BII speed. Betamax had now lost its highest-quality contender and left those early adopters of the format in need of a new machine. But would they stick with Beta or opt for the format that was gaining in popularity—VHS?

WHY I LOVE IT

Since it shares the design of the OG Betamax—the SL-7200—and is capable of recording in multiple speeds, the SL-8200 is my favorite VCR of all time. It is an absolutely gorgeous machine with the ability to play all of the prerecorded movies that would be released later, since BII was used. In my opinion, you just cannot beat that classic aesthetic with the playback capabilities needed for modern-day Beta viewing.

TENNIS
STARMASTER
SLOT RACERS
ASTEROIDS
COMBAT
12 38
SONY
SONY
K-30
SONY
VIDEO CASSETTE
K-60
SONY

1978 SONY BETAMAX SL-3000

Format: Beta **Weight:** 20lb 1oz (9.1kg) **Introductory price:** $1,299.95 USD

In the early days of the VCR, a big draw for many video enthusiasts and families was the ability to connect a video camera and record memorable moments straight to a videocassette. Unlike the previous generations of film cameras, such as 8mm, there was no need to develop the footage. Immediately after recording, the videocassette could be rewound and watched. A drawback, however, was the need to plug the video camera into the VCR, which made it impossible to capture events that took place further from your VCR than the cable could stretch.

To accommodate those who wanted the option of recording outside of the home, the portable VCR was introduced. The first portable Betamax for Sony was the SL-3000. In order to be portable, all unnecessary weight in the form of a tuner and timer was removed, leaving just the manual controls needed for a VCR. Sony had also eliminated the ability to record in the highest quality (BI) by this time. The SL-3000 could playback BI recordings but could not record them. Since the market implied that the public was more interested in recording duration than quality, Sony had transitioned BI to being a professional format. It was just too good for the average Joe, and BII was the only option.

The SL-3000 was a battery-powered recorder hung by a strap on the user's shoulder while footage came from a hand-held corded camera. The video camera (HVC-1000) was a separate purchase for $1,400! While very expensive, the color video camera was an emerging technology and most options on the market were still black and white. Even though it was a remarkable advancement for video cameras, at over 20lb, the HVC-1000 was not a comfortable experience. That did not prevent dads from carrying them around on family vacations or capturing their kid's recital. A little shoulder pain is a small price to pay for a physical copy of lasting memories.

If buying the VCR and video camera was not enough, in order to use the portable VCR to record your favorite show at home, an AC adapter, the AC-300C, and a separate tuner, the TT-3000, could also be purchased.

WHY I LOVE IT

A portable Betamax that could be used with a camera to record anything you could imagine—including your very own movies—is magical! After recording, just hook it up to the closest TV to see how your footage came out. There was a good chance that the lighting looked terrible, but it was better than nothing.

SONY
Betamax
STOP
PAUSE
EJECT
REW
PLAY
F FWD
REC
AUDIO DUB
PAUSE
STAND BY
PORTABLE
VIDEO CASSETTE
RECORDER
BATTERY
CAMERA/TV
MIC
EARPHONE

RACKING
RF OUT
DC IN
SEE INSTRUCTION-MANUAL
VIDEO OUT
LINE OUT

1979 JVC VIDSTAR HR-6700U

Format: VHS **Weight:** 31lb (14kg) **Introductory price:** $1,350 USD

With this second-generation Vidstar, JVC made further advances with the VHS line by offering a new, even faster speed known as Extended Play (EP) or Super Long Play (SLP) that allowed for six hours of recording on a T-120 cassette. The HR-6700U was the first VHS machine with four video heads. One set of 58.5-micron heads intended for SP recordings, and one set of 19.5-micron heads used for EP recordings. Recording at the LP speed was not available on this Vidstar, and many VHS brands were in the process of eliminating it. The quality of recordings by the HR-6700U in SP mode was the best available from any consumer VCR at the time, including Beta machines. At this setting, however, only two hours of recording was possible. Most people preferred to use the six-hour EP mode that allowed them to record more programs on each tape.

The HR-6700U also offered an improvement on the programmable recording features. It used a microcomputer to enable six channels to be recorded at different times over a seven-day period. Programming the recording day was no longer the struggle it had been on the earlier RCA VCT-400. You could simply select the desired day of the week. In EP mode, you could fit multiple movies or sporting events on the same tape.

During playback, the Vidstar allowed for still-frame viewing, slow-speed viewing, or speed play, which played back at twice the normal speed in SP mode, or three times normal speed in EP mode. All of these features could also be accessed using the six-function wired remote. Previous remotes we have discussed were just a pause button used to eliminate commercials during recording, but now remote controls that offered multiple functions were being produced.

WHY I LOVE IT

Although not as aesthetically pleasing to me as the first-generation Vidstar models, such as the HR-3300, the HR-6700U is still a beautiful machine with many awesome features. When it comes to VHS VCRs, I always lean toward the JVC Vidstars as the leading format in terms of style combined with functionality. You will notice that each of the Vidstars has minimal wood grain on them, unlike almost every other VCR of the time. JVC was pushing the technology to appear sleeker, futuristic and more exciting.

1980 SONY BETAMAX SL-5800

Format: Beta **Weight:** 36lb (16.3kg) **Introductory price:** $1,450 USD

By this time, both Beta and VHS had created a third speed, allowing even more programming per tape. For Beta, it was designated as BIII and allowed for up to five hours of recording time on an L-830 tape. This was a huge difference from the one-hour that was initially possible on the original Betamax.

These longer play times were not only made possible by changing the tape speed through the VCRs, the thickness of the tape also had an impact. Thinner dimensions allow for more tape to fit in a cassette. For Beta's L-750 tape, the thickness was decreased to 15 microns, which was decreased again to 13 microns for the L-830 tapes. But a decrease in thickness made the tape more prone to damage. VHS manufacturers followed suit and in addition to their standard 20-micron T-120 tape, they began producing a thinner T-160 tape to maximize content. The T-160 allowed for a whopping eight hours of recording time! This was a major factor in the war between VHS and Beta, and, according to the December 1980 issue of *Video* magazine, the ratio of consumer VCRs was split about 60 percent VHS to 40 percent Betamax at the time.

The Sony SL-5800 was given the name "The Time Commander," because it gave the viewer full control of the speed of the image they were watching with variable-speed picture search. Using a wired remote control, one could easily scan through a video at speeds ranging from frame-by-frame to full fast-forward while seeing the image on screen the entire time. This was possible by adjusting a knob on the remote control that controlled the speed of the tape, while selecting to progress the video in forward or reverse direction. It was a great advancement in the search capabilities of home VCRs. The previously released SL-5400 and SL-5600 introduced visible picture search while fast-forwarding. This feature was not yet available with VHS machines since the tape had to be unthreaded from the video heads.

WHY I LOVE IT

Even today, the variable-speed picture features of the SL-5800 are highly impressive. I am sure that at the time of its introduction, it was a marvel to behold. Combine that with the optional AG-300 (see page 102), and you have one of the coolest Betamax setups around!

Beta
HULKAMANIA
It's Alive
VACATION
SONY
TIME COMMANDER SL-5800

SONY
AG-300
L-750HX
TIME COMMANDER SL-5800
8:29
SONY
CASSETTE AUTO CHANGER AG-300
SONY

ITEMS OF INTEREST 08

The SL-5400 was only capable of recording one event across three days, while the SL-5600 and SL-5800 could record four programs across 14 days. All three of these machines were compatible with the AG-300 automatic cassette changer called the "Betastack." This was a truly spectacular piece of equipment that Sony hoped would quiet naysayers who remarked that VHS could record more than Beta. The Betastack held four Beta tapes and could cycle through them, recording up to an amazing 20 hours of programming without assistance. It could even be programmed to switch tapes after recording a complete event to prevent a program from being split over multiple tapes. If multiple cassettes were needed for a single program, the AG-300 could swap tapes in just 15 seconds, resulting in minimal loss of content. The obvious downside is that multiple cassettes were being used. A Beta tape was still not capable of the recording length of a VHS one. However, the Betastack was still potentially useful. Not only was it capable of recording on multiple tapes, but it could be used to play them as well. Imagine a movie marathon where your films of choice were loaded up and, as one movie ended, the next started playing automatically. That is my kind of movie night!

1

1980 SONY BETAMAX SL-5600

The SL-5600 improved on the SL-5400 by increasing the number of recording events from one over three days, to four spanning a whole two weeks!

2

1980 SONY BETAMAX SL-5800

Known as the "Time Commander," the SL-5800 (see page 100) took Betascan to a new level by giving the viewer full control over the speed during forward and reverse viewing.

3

1979 SONY BETAMAX SL-5400

The SL-5400 introduced a ground-breaking feature called Betascan, which allowed the user to visibly search a tape during rewind and fast-forward.

9:48
TIME COMMANDER SL-5800
SONY
9:48
SONY
1
BetaScan
9:48
SONY
SONY
VIDEO CASSETTE
L-500
L-500
SONY
Dynamicron
L-750
SONY Dynamicron L-750
L-830
SONY Dynamicron L-830

1983 JVC VIDSTAR HR-7100U

Format: VHS **Weight:** 21lb 10oz (9.5kg)

JVC continued its Vidstar line with the HR-7100U. Although not as sleek as the previous entries, the HR-7100U has a fun design. The large, colorful buttons capture your attention and help make the Vidstar easy to use. By this time, video rental stores were booming. Most of them were still mom-and-pop stores, or sections of grocery stores or gas stations; few big chains had entered the market. Perhaps the best-known chain, Blockbuster Video, would not open for a couple more years. Even still, JVC and other VHS brands were making their VCRs more intuitive to operate. The days of needing to be a videophile with an engineering degree to set up and use video equipment were fading. The JVC HR-7100U was a step in the user-friendly direction. The functions were clear enough for anyone to get family movie night started.

Surprisingly, JVC was still producing top-load VCRs, even though the first front-load machine, the Sharp VC-6800, had been introduced back in 1980. I am not complaining, because I actually prefer the top-loaders. The tape popping up upon eject is super-satisfying. Sure, there are many benefits to front-loaders, but it is hard to beat the aesthetics of those early VCRs.

Another note I must make about the HR-7100U is that it is perhaps most recognizable as the VCR seen in the intro to the ABC sitcom *The Goldbergs*, which aired between 2013 and 2023. The show is about a family in the Eighties, and the youngest son is obsessed with TV and movies, and makes home movies with a video camera.

WHY I LOVE IT

By now you know that I love the JVC Vidstar line, and what is not to love about those big, beautiful, colorful buttons? They stand out so much and make you want to press them. It is hard to resist popping in a movie when you have this machine around.

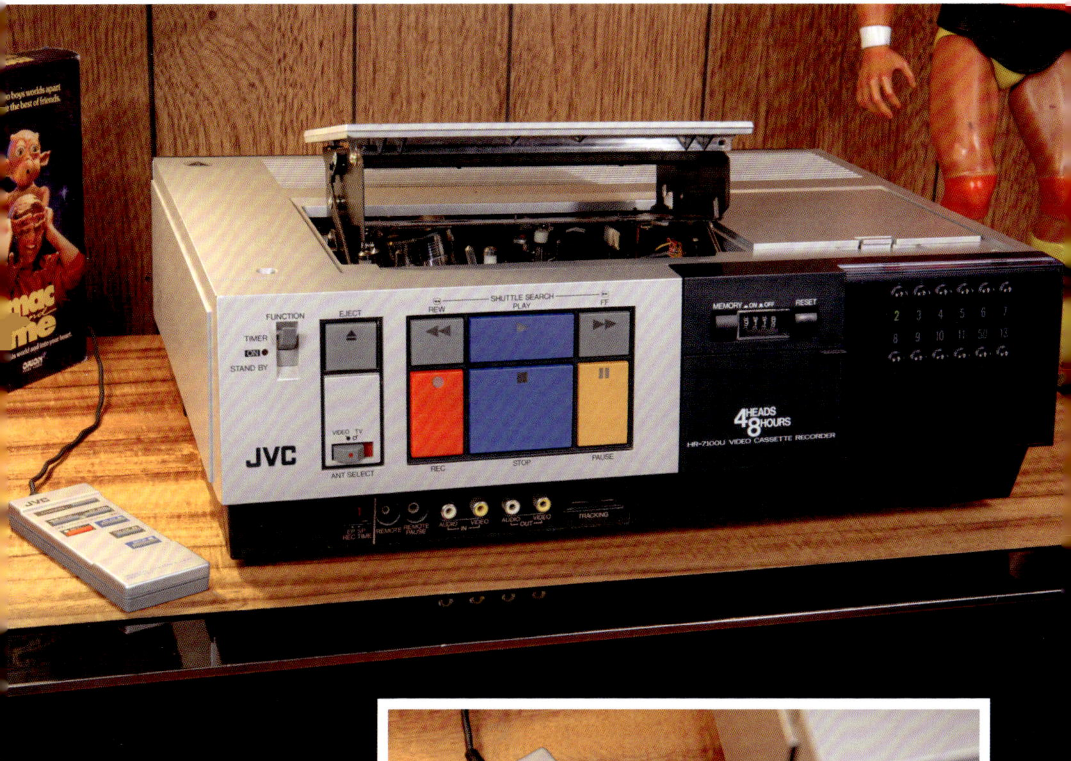
JVC
FUNCTION
TIMER
ON
STAND BY
EJECT
SHUTTLE SEARCH
REW
PLAY
FF
REC
STOP
PAUSE
VIDEO TV
ANT SELECT
MEMORY ON OFF
RESET
4 HEADS 8 HOURS
HR-7100U VIDEO CASSETTE RECORDER
REMOTE
REMOTE PAUSE
AUDIO VIDEO IN
AUDIO VIDEO OUT
TRACKING

JVC
CHANNEL
REW
FF
SHUTTLE SEARCH
PLAY
REC
PAUSE
STOP
REMOTE CONTROL UNIT
JVC

1983 SONY BETAMAX SL-2710

Format: Beta **Weight:** About 20lb (9kg) **Introductory price:** $1,300 USD

Although nearly six million Americans had already purchased a VCR by 1983, demand was not as high as the Beta and VHS companies had predicted. The market was overwhelmed by a surplus of VCRs, causing stores to begin slashing prices to get rid of excess stock. Competition and lower prices proved to be successful ways to get more VCRs into consumers' homes. VHS was outselling Beta around the world. In the US, three times as many VHS VCRs were sold compared to Beta machines. At the same time, more movies were being released on videocassette, and many retail and video stores did not have the room or desire to stock the same content on multiple formats. It was becoming obvious that more customers preferred VHS, and this resulted in many stores dropping Beta copies from their inventory—another huge blow to the Beta community. Even if people had, or were interested in getting, a Betamax, not having access to new releases made it less likely that they would invest in the format.

By this time, over 2.5 million Beta players were in use, but VHS was still outpacing Beta considerably. Zenith jumped ship and began selling VHS machines. Philips in Europe also made the move to VHS, indicating that VHS was now dominant around the world.

Still in the fight, Sony introduced Beta Hi-Fi to bring the best audio quality the home videocassette had yet produced. Beta Hi-Fi had significantly better sound quality than the contemporary stereo audio on VHS, and gave poor Beta sales a boost. The first Hi-Fi VCR was the Sony SL-5200 Betamax, which predates the SL-2710. Many prerecorded Beta tapes produced after this point would have the Beta Hi-Fi stamp on the cover, to ensure consumers knew they were getting higher-quality sound. It would take VHS another year or so before catching up and offering its own hi-fi audio.

WHY I LOVE IT

I found this SL-2710 at the same estate sale where I picked up the Sony Trinitron KX-2501A. It was also in the garage and coated in a thick layer of dust. It looked like it was destined for the dump, but I couldn't let that happen. After a good cleaning and belt replacement, the Betamax looked and functioned like new. The fact that it's an early Beta Hi-Fi unit, with a super sleek silver style, makes it one of the more impressive Beta players from the early Eighties

CHANNEL
AM PM
SUN MON TUE
WED THU FRI
SAT NEXT WK EVERY DAY
THIS WK
1 2 3 4 5 6 7 8 9 0 C
REMOTE SENSOR
PROGRAM SELECT
ON VTR OFF-TV
WEEK AFTER NEXT
AUX
RECORD
TIMER
NEXT
ON/OFF
TIMER REC
EVENT DAY TURN ON OFF CH
TIMER SET
QUICK TIMER
CHECK
BACK
SONY
STEREO
L
R
CLOSE

SONY
Betamax
POWER
STEREO VIDEO CASSETTE RECORDER SL-2710
PLAY
REW
FF
STOP
PAUSE
LEVEL
REC LEVEL
INPUT SELECT
TRACKING
NEXT
SONY
CHECK
BACK

1986 SONY BETAMAX SL-HFT7

Format: Beta **Weight:** 28lb (12.7kg) **Introductory price:** $1,000 USD

Soon after the release of the SL-2710, which was an early Hi-Fi model, the letters "HF" were added to model numbers of hi-fi-capable Betamax machines. Some lower-level models had the letters "HFR," indicating that they were hi-fi-ready, but in order to utilize this function, users had to purchase a separate Beta Hi-Fi stereo processor and hook it up to the machine. This SL-HFT7 is hi-fi-capable; the "T" indicates that this is a theater model.

SuperBeta Hi-Fi Theater models were designed to bring the theater experience into your living room by incorporating great audio and superb visuals. The sound came from the Beta Hi-Fi feature, while the visuals were due to a new advancement called "SuperBeta." Sony pushed the Beta technology to improve luminance, which resulted in an image with finer detail. According to Sony, the picture quality was up to 20 percent sharper and increased the horizontal resolution to 290 lines.

One of the features that put "Theater" in the name of this Betamax was the ability to change the audio mode. There are buttons on the top-right corner of the player with options to change the sound system between "Movie Theater," "Concert Hall," and "Simulated Stereo." The "Movie Theater" option was selected when watching a hi-fi movie, "Concert Hall" was appropriate for music videos or concerts on tape, and "Simulated Stereo" may be the best option for movies not recorded in hi-fi/stereo.

WHY I LOVE IT

This is one of my more frequently used machines. It is a great-looking VCR, and the white just makes it pop against a typical black or wood-grain TV. It has all the features needed to playback prerecorded movies.

THE TALISMAN
THiNG
STEREO VIDEO CASSETTE RECORDER SL-HFT7
STEREOCAST
STEREO
SAP
Beta Hi-Fi
THEATER SOUND SYSTEM
MOVIE THEATER
CONCERT HALL
SIMULATED STEREO
PLAY
PAUSE
CHANNEL/TIMER
VOLUME
MUTING
SONY
BASS BOOST
THEATER LEVEL
BALANCE
REC LEVEL

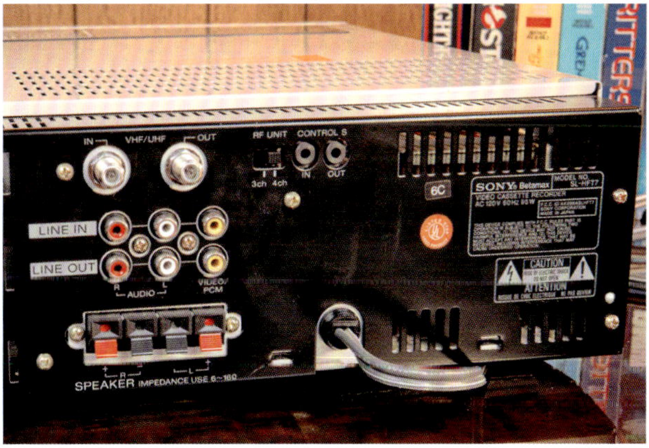
IN
VHF/UHF
OUT
RF UNIT
3ch 4ch
CONTROL S
IN
OUT
LINE IN
LINE OUT
R
L
AUDIO
VIDEO/PCM
SPEAKER IMPEDANCE USE 6~16Ω
SONY Betamax
VIDEO CASSETTE RECORDER
MODEL NO. SL-HFT7
CAUTION
ATTENTION

1986 PANASONIC OMNIVISION PV-1566

Format: VHS **Weight:** 13lb 6oz (6.1kg)

In response to the improved quality of Sony's SuperBeta, JVC and other VHS brands achieved a sharper image with HQ VHS. HQ stands for "High Quality," and this comes from a reduction in noise, color streaking, and jittery edges. Unlike SuperBeta, there was no increase in horizontal resolution. HQ VHS was still 240 lines, but recorded images did have noticeably sharper edges, resulting in a better picture.

This Panasonic PV-1566 is an example of an early HQ VHS machine. It has four-heads and is capable of recording in SP, LP, and SLP speeds, although the SP and SLP speeds are best for using the special effects during playback. These special effects include playing back a recording at 2x speed and using the "slow" button to advance through a recording frame-by-frame.

"For decades, being able to pop a new release into your VCR was a highlight of many Friday nights."

WHY I LOVE IT

Although we have entered the era in which many entry-level VCRs were losing their style in order to make them more affordable, this one still looks good. It is a relatively simple machine, but the clear door with its blue Panasonic logo is a nice touch. I found this VCR at an estate sale for about $5, and I will pick up these kinds of machines at that price all day long, whether they work or not.

NIGHT OF THE CREEPS
THEY LIVE
BLOB
Panasonic
Panasonic OmniVision VHS
HQ
EJECT
Hi-Fi
power
pause/still
stop
slow/f. adv
play/×2
rec
Hi-Fi Audio

Panasonic
Panasonic OmniVision VHS
HQ
VCR/TV
level meter
Hi-Fi
EJECT
off on HD track
power
pause/still
stop
slow/f. adv

1986 SONY BETAMAX SL-HF750

Format: Beta **Weight:** 24lb 3oz (11kg) **Introductory price:** $1,300 USD

As Sony saw the writing on the wall indicating the end of Beta, the company went all-in on making Betamax the ultimate VCR for those videophiles who still supported it. This can be seen in the SL-HF750, which is one of the most unique Betamax models ever released. One of the physical characteristics that stands out the most is the automatic tape tray that extends and pops out for an extravagant loading and unloading experience. But the real surprise is that Sony brought back BI recording, after almost a decade, the recording speed that was reserved for the professional Betamax had finally returned to consumer models. Now, I wish they had done this much sooner or never gotten rid of it in the first place, but the fact that the higher-quality recording capability was back was a big deal. Unfortunately, it was included only on the higher-end models, aimed at those who prioritized higher-quality recording over longer record times or smaller price tags. Of course, the machine was still able to record in BII, BIII, Beta Hi-Fi, and SuperBeta as well as Super Hi-Band BIs (5.6 MHz), which incorporated the SuperBeta functionality into the BI speed for an even more improved picture. It was capable of pretty much anything Sony had developed for the Betamax up to that point.

Now, back to that automatic tape tray. It's not just for loading and unloading a cassette. There is also an open/close button that extends the compartment from the VCR while the tape is playing and allows you to see the tape in action, or check how much tape is left while recording. The way that the entire cassette compartment extended from the Betamax was referred to as linear-skating. Paired with the large number of features and editing capabilities, this is essentially a professional machine marketed to the home-video enthusiast.

WHY I LOVE IT

The SL-HF750 is an absolutely beautiful and unique VCR. Seeing this machine's linear-skate feature in action, when it is essentially reaching out to accept or deliver a Beta tape, is delightful. The moving parts are what I like most about these old analog machines, and the SL-HF750 does not disappoint.

LINEAR SKATE CASSETTE LOADING
SONY
OPEN/CLOSE
POWER
ON/OFF
LEVEL
MIN
MAX
HEADPHONES
EJECT
STEREO VIDEO CASSETTE RECORDER SL-HF750
REMOTE
SENSOR
REC LEVEL
TV/VTR
COUNTER RESET
TAPE RETURN
PEAK PROG METER
STEREOCAST
Super Beta
REV

POWER
ON/OFF
EJECT
REMOTE
SENSOR
LEVEL
HEADPHONES
STEREOCAST
Super Beta
REV
SONY

1986 SONY BETAMAX SL-HF1000

Format: Beta **Weight:** 30lb 4oz (13.7kg) **Introductory price:** $1,700 USD

Soon after the SL-HF750 hit the market, Sony introduced what many consider to be the greatest Betamax ever made—the SL-HF1000. By this time, Beta's share of VCR sales had dropped to well below 20 percent. But instead of going out with a whimper, Sony was steady, swinging to stay in the fight. However, with machines such as the HF750 and the HF1000, the company was not targeting casual viewers, like the family in need of a player to watch a movie rented from the local video store on a Friday night. It was targeting the home-video enthusiasts who planned to record frequently and wanted professional-level editing capabilities. You have probably heard about, or may even remember, how many people had VCRs whose clocks constantly blinked "12:00," because no one knew how to set the correct time. If people couldn't set the clock, they certainly weren't setting up multiple recording events without help. The HF1000 was not the machine for this kind of user. Sony was catering to the videophiles who expected a lot from their VCR.

Some of the features of the SL-HF1000 include recording and playback capabilities of BIs, BII, BIII, Beta Hi-Fi, and SuperBeta. The BIs was further improved from the SL-HF750 to 6.0 MHz Super Hi-Band. The SL-HF1000 had twin flying erase heads that allowed for glitch-free editing and a built-in character generator for adding text to the recordings. A programmable timer was able to handle up to eight recording events spanning 21 days. It also came with the Commander wireless remote control, which gave you the freedom to access practically every function of the VCR from the comfort of your couch. For less searching of the more common controls, a tinted cover eliminated unwanted access to a multitude of rarely used buttons. One of the coolest controls comes in the form of the jog-and-shuttle dial that is found on both the deck and the remote. The similar and more primitive dial on the wired remote of the SL-5800 harkens back to the earlier days of Betamax. The shuttle ring allowed for continuous bidirectional

RESTRICTED
POWER
ON/OFF
EJECT
Super Beta
LEVEL
HEADPHONES
SONY
Super Hi-Band
6.0MHz
STEREO VIDEO CASSETTE RECORDER SL-HF1000

SONY

POWER
SONY
EJECT
ENTER
CLEAR
JOG-SHUTTLE
REC
PAUSE
PLAY
FF
REW
STOP
VTR/TV
RMT-148

tape speed of 1/5x, 1x, and 2x, while the jog dial gave you the ability to scroll through tape at variable speeds.

There are dozens more features I could bring up about this Betamax, but let's not turn this into a sales brochure. I just want to stress the effort and attention Sony put into it. We now know that it was too little too late, but Sony fought to keep Betamax alive at a time when it was also helping usher in a new, smaller format called Video8, and just before it made the decision to start the production of its own VHS machines.

"With the HF1000, Sony was targeting the home-video enthusiasts who planned to record frequently and wanted professional-level editing capabilities."

WHY I LOVE IT

This is a truly remarkable VCR. It has way more features than the average person would ever use, but if you want to do a lot of editing this really is the machine for you. If you are just watching prerecorded movies, this deck is way more than you will need, and playback of a prerecorded movie will look essentially the same on any of the machines from the same era. The real advantage here is in the potential for home recordings.

SONY
POWER
REVERSE
FORWARD
NEXT
MARK

VHF/UHF
CONTROL
CONTROL S
CAMERA REMOTE
LINE IN 1
MONITOR OUT
LINE OUT
SL-HF1000

ITEMS OF INTEREST 09

1

1985 SANYO BETACORD

Similar to the way Sony called its Beta players "Betamax," Sanyo referred to its machines as "Betacord." I purchased this one from an older gentleman who was selling his player and collection of recordings. His tapes are very well labeled and numbered, and came with a handwritten catalog listing the movies on each tape.

2

1987 ZENITH VRD530HF

In a large stack of VCRs, this vertical example by Zenith (see page 120) really stands out. There are many examples of both front- and top-load machines, but a vertical VCR that loads the tape sideways is hard to come by.

3

1984 PANASONIC PV-1720

Released in 1984, this Panasonic machine is one of the first VHS VCRs capable of stereo audio. It can even record in SP, LP, and SLP speeds.

I'm here for the
VCRs
SONY
Betamax
VIDEO CASSETTE
K-60
THE KARATE KID
GHOSTBUSTERS
PIECES
TIME COMMANDER SL-5800
K-30
L-500
MASSACRE
The MONSTER SQUAD
WAYNE'S WORLD
Happy Gilmore
PULP FICTION
SONY
SANYO
1
SANYO
JVC
2
JVC
Panasonic OmniVision VHS
3
STEREO Video Cassette Recorder VHS
JVC
4 HEADS 8 HOURS
HR-7100U VIDEO CASSETTE RECORDER

1987 ZENITH VRD530HF

Format: VHS **Weight:** 16lb 10oz (7.3kg) **Introductory price:** $900 USD

In 1987, JVC introduced Super-VHS (S-VHS), which further improved VHS picture quality. Horizontal resolution was pushed to 400 lines, which was about 30 percent more than television broadcasts, and twice as good as standard VHS. The first S-VHS to hit the market was the JVC HR-S7000U, which sadly I have not added to my collection yet.

The introduction of the S-VHS also popularized the S-Video input—and, no, the "S" in S-Video does not stand for "Super." S-Video improved image quality to the monitor by sending separate signals for both the luminance and the chrominance (the two components of a colored image) directly from the source. Therefore, the "S" in S-video stands for "separate," since those video signals had been separated. It did not require the signal to go through a tuner like RF (radio frequency), and did not combine the signals like AV (audio/visual). This allowed for cleaner, more detailed signals.

The S-VHS machines were backward-compatible with earlier types of VHS tapes, but you could not play an S-VHS recording on a standard VHS VCR. Special tapes, such as the ST-120, were needed to capture the finer detail of the S-VHS. These tapes had a "recognition hole" on the bottom-right side of the cassette for the VCR to distinguish them from a standard VHS.

Even though S-VHS was a major improvement in the VHS line, it didn't gain enough popularity for studios to release movies at this higher quality. Again, the common VCR user had no interest in spending extra money on this newer technology. After all, they were still reckoning with the recent release of the HQ VHS as seen with this one.

Zenith had never been afraid to try new and interesting design choices, and in 1987 it released one of the least VCR-looking VCRs ever sold. The VRD530HF is a vertical, side-load VCR in which the cassette tape is inserted longways. It looks more like a futuristic (at that time) computer tower than a VCR and is more compact too. In some cases, it was possible to just slide it in next to a TV, instead of requiring an entire shelf below the TV, as a typical VCR would.

While this VCR does not contain the latest S-VHS technology, it was still a good-quality HQ VHS machine with many features. Its four-head hi-fi deck could record up to eight programs spanning two weeks. This Zenith records in SP and EP with 240 lines of horizontal resolution but plays back all three speeds (SP, LP, and EP). All the more commonly used buttons are within easy reach, with the less-used buttons hidden behind a swing-down panel, which keeps this machine looking sleek.

WHY I LOVE IT

It's not common to see a vertical VCR, and a machine that also loads the tapes in sideways is a sight to behold. The sleek combination of matte and glossy black with the large clock display screams late-Eighties. Zenith knocked it out of the park with this design.

ZENITH
STEREO AUDIO/VIDEO RECORDER
INDEX
MARK
ERASE
TV/VIDEO
CLOCK/CNT/REM
COUNT RESET
CHANNEL
DOWN
UP
PROG
AM/PM
SKIP/CNCL
CLOCK ADJ
ENT
GO-TO
POWER
EJECT
PLAY
STOP
PAUSE
RECORD
SLO-MO
REW
F FWD
SPEED
SEARCH
PICTURE SHARPNESS
TIMER
TRACKING
AUTOMATIC TIMER RECORDING

1988 SONY SLV-50

Format: VHS **Weight:** 14lb 10oz (6.3kg) **Introductory price:** $600 USD

It finally happened! After many years of proclaiming it would never produce VHS VCRs, Sony gave in and listened to the desires of the masses. The first VHS machines the company released were the monaural SLV-50 and the SLV-70HF, which had hi-fi audio, as indicated by the "HF" in the model number. The model shown here is the SLV-50. Both models were HQ VCRs and gave the user some significant editing features, such as a flying erase head and the ability to hook two machines together to program and edit footage from one tape onto another. It also enabled users to record up to eight programs over the span of a month.

Although these VCRs offer many good features, they look as if Sony just produced them in order to appease consumers. They certainly do not have the soul and aesthetic of the company's Betamax machines. Sony eventually made much nicer-looking VHS decks and, not surprisingly, this led to sales of higher-end VHS machines in the following years. Sony was now in the consumer market with Beta, VHS, and Video8.

Do not make the mistake of thinking Sony was done with Beta at this point. In the same year it both further improved the image of the Betamax and released Extended Definition (ED) Beta. The ED Beta came hot on the heels of S-VHS and emphasized Beta's continuing superior quality with a whopping 500 lines of horizontal resolution. While the machines were capable of recording glorious images, no prerecorded movies were being released at such high quality. Anyway, these machines were not intended for the ordinary consumer. Just a couple of ED Beta models were released in North America, with price tags approaching $3,000 USD, which made them out of reach for most buyers.

WHY I LOVE IT

Although you can probably tell I slightly favor Beta over VHS, I am still a VHS fan and collector. I'm also a Sony fan and collector, so it's nice to have a Sony option when watching VHS. But I'll be honest, I always lean toward JVC for my VHS needs.

SONY
REC MODE
SP/EP
SHARPNESS
SOFT SHARP
TRACKING
NORMAL
SLOW
EJECT
ADD
ERASE
PRESET
PRESET TUNING
CLOCK SET
CLEAR
CHECK
NEXT
ON/OFF

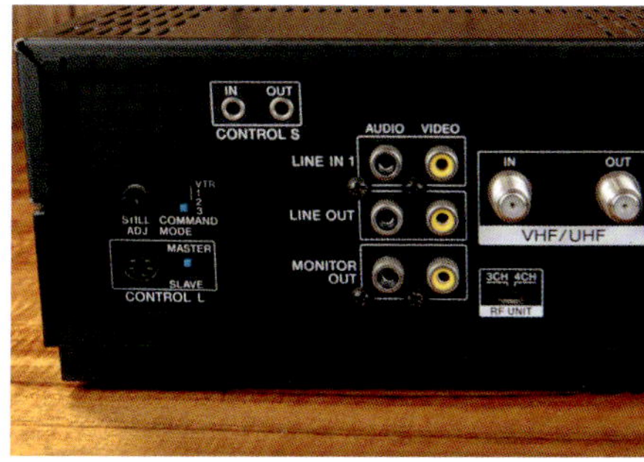
IN OUT
CONTROL S
AUDIO VIDEO
LINE IN 1
LINE OUT
MONITOR OUT
IN OUT
VHF/UHF
STILL ADJ
COMMAND MODE
VTR
MASTER
SLAVE
CONTROL L
3CH 4CH
RF UNIT

VHS
HIGH QUALITY PICTURE/FLYING ERASE HEAD/DA PRO 4 HEAD/DATA SCREEN
Digital Picture
EJECT
SONY
EDIT MONITOR
SYNC EDIT
INSERT
POWER
ON/OFF
HQ
LINE IN 2
VIDEO AUDIO
TV/VTR
COUNTER RESET
CHANNEL
INPUT SELECT
VIDEO CASSETTE RECORDER SLV-50
THE TERMINATOR
WEIRD SCIENCE
STALLONE
OVER THE TOP

1991 SONY BETAMAX SL-HF2100

Format: Beta **Weight:** 24lb 5oz (11.1kg) **Introductory price:** $1,699 USD

In 1990, Sony released a 15th anniversary Betamax in Japan. The machine was not intended for release in North America, but even though the Beta format was fading out, Betaphiles demanded that Sony release it in the US. The following year, the SL-HF2100 arrived. This is the most outlandish and ambitious Betamax (and possibly VCR of any type) ever created. Imagine this—after more than a decade and a half of VCRs covered in buttons, levers, toggles, switches, and knobs, Sony released a model with no physical buttons! It had a sleek deck with a clear front panel that folds down to reveal the cassette door. An amber backlit touch panel reveals the location of the touch functions. Next to the cassette insert is a display showing the clock, the type of tape or recording method in progress, and a status wheel that spins to indicate the direction and speed in which the tape is moving. The clock and status wheel are still visible when the front panel is closed.

It doesn't stop there—even the remote control is an LCD touch panel. Only the common controls such as play, stop, and rewind are initially displayed on the remote, but as you use the controls, the screen changes to give you access to more functions. This remote is also the only way to set the clock on the VCR: you set the time on the remote and then transmit it to the deck. Both the remote and the deck emit a beep to indicate that transmission was successful as the time is updated on the deck. This had to feel very futuristic at the time of release.

Much like the SL-HF1000, this Betamax is capable of playing and recording practically every speed Sony had developed for the format, with one exception—it was not compatible with ED Beta, which is unfortunate. An improvement over the SL-HF1000 are the Opt-4 video heads. Instead of using two different video heads that specialized in separate recording speeds, the SL-HF2100 has a wider pair for record and a narrower pair for playback, which improved picture clarity.

In 1993, Sony released a stripped-down version labeled the SL-HF2000. Physical buttons replaced the touch screen, as the design elements that made the HF2100 such a marvel were removed. The HF2000 would turn out to be the last Betamax produced for the US market.

WHY I LOVE IT

While it's not quite the last Betamax, you can see Sony was willing to push the limits for true fans of the line. This is a spectacular machine that needs to be seen to be appreciated. I'm still in shock that I found this one in a Value Village thrift store in Everett, Washington. This was the only time I had ever been to that particular store, and I figured I would check the glass cases on the way out. There to my surprise was one of the best Betamaxes ever released. I knew I wasn't leaving the store without it.

6.0MHz Super Hi-Band
SONY

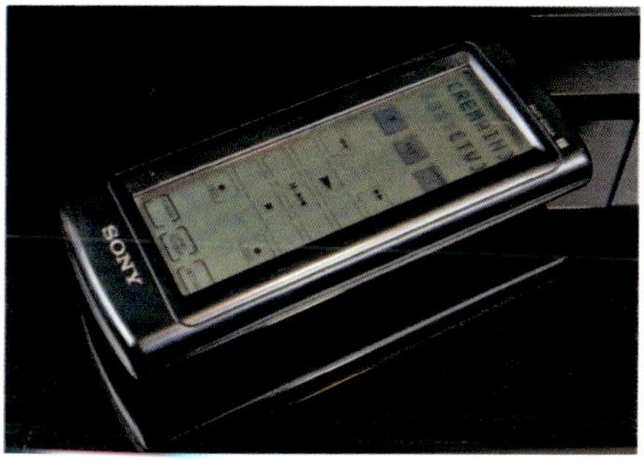
SONY

SONY

04

HALLOWEEN II
ALL NEW
From The People Who Brought You "HALLOWEEN"..
More Of The Night He Came Home.
HALLOWEEN III

VIDEODISCS: LASER VS CED

Just as the war for home-VCR supremacy was getting started, a completely different type of format was about to enter the market—the videodisc. In general, a videodisc is simply a disc that has a video signal stored on it. Different methodologies are used to store video information on these discs, and in this section we will investigate the method used by the LaserDisc and Capacitance Electronic Disc (CED) formats.

Although LaserDisc and CED are both videodiscs, that is pretty much where the similarities end. They are two entirely different formats. LaserDiscs use a highly focused laser to read millions of pits in concentric circles within a central layer of the disc. As the disc spins within the machine, the laser is projected at the pits and then disrupted and reflected back into the machine through a series of mirrors and prisms, before being decoded into a picture signal. CED videodiscs incorporate a diamond-tipped stylus that makes physical contact with the disc and measures the capacitance along grooves in it. This electrical measurement is then converted into video and audio signal.

In the previous section, we explored the videotape format war by looking at how VCRs improved and evolved as the battle was waged. Similarly, here we will look at particular videodisc players and discuss how the formats were received by the general public. You may wonder why I didn't group these machines with VCRs and compile everything into one big section about the format wars. That's because the videodisc was so wildly different that it really was not in direct competition with the videocassette. Although all these formats are primarily used by modern-day collectors to watch prerecorded movies, you have to keep in mind that they were originally marketed to meet different consumer needs. As previously discussed, the big draw of the VCR was the ability to record programming from the TV and make home videos in conjunction with a video camera. Neither of the videodisc formats had these capabilities, and, therefore, really had no chance of dethroning the videocassette. Videodiscs did have their own advances, which made them worthy of being released. So, let's load the discs and get them spinning.

1978 MAGNAVOX MAGNAVISION 8000

Format: LaserDisc **Weight:** 40lb (18.1kg) **Introductory price:** $500 USD

In 1972, MCA gave the first public demonstration of the laser videodisc, a revolutionary technology that was sure to transform media storage and the way people watched movies at home. Unfortunately, it would not come to the market until 1978, and by this time both Beta and VHS had been infiltrating households for several years. These could not only play prerecorded movies, but also record programs straight from the TV.

After many years of promises and buildup, the first videodisc system to hit the US consumer market—Magnavox Magnavision developed by Philips and MCA—arrived in 1978. By using optical technology rather than relying on physical interaction between the media and the player, the Magnavision ensured that discs could be played thousands of times without image degradation. This system used a laser to read pits in the disc and translate the information into the images and sound that were output to the television. Although the media for this system would later become known as "LaserDiscs" (a term developed by Pioneer), early on they were referred to simply as videodiscs. The introduction of additional, incompatible videodisc systems such as the RCA CED system prompted the need for clarifying terminology. Different companies also had their own terminology, such as "DiscoVision" and "LaserVision."

One of the key features that set the Magnavision apart from videocassette recorders was the ability to scan forward and reverse at variable speeds while producing a crystal-clear image. Freeze-frame was also available, and the user could cycle through, frame-by-frame, without any damage to the player, since there was

MAGNAVOX
PLAY
FAST
AUDIO
SEARCH
INDEX

no physical contact between the disc and the reader. The discs were also able to store 54,000 images or frames per side, which made the format a good candidate for storing media such as documents or educational materials.

A common misconception is that these systems did not become mainstream because they were too expensive, but at $500, the player was initially priced well below a VCR, which cost about $1,000 on average. At the start, the discs ranged in price from $2 to $10, which was also well below the price of videocassettes. Unfortunately, due to poor sales and challenges with the reliability of discs and players, these prices began to increase.

Magnavox also encountered many difficulties in getting the players built and distributed, which meant that customers had a hard time finding them. They were initially only available in Atlanta, Georgia, before slowly spreading to other cities. To make matters worse, finding an adequate selection of discs was no easy task, and if you did find the disc you wanted, it was not guaranteed to work. Production problems meant many of the early discs had quality-control issues that made them unwatchable. Customers would frequently exchange a bad disc for another copy that did not work any better. Due to these struggles, manufacturers raised the prices of both the systems and discs to account for all the losses. By 1980, the player had increased to $775, while the price of discs more than doubled to between $5.95 and $24.95.

A major downside of the videodisc was that it was not recordable. This may not sound like much of a problem to us now, but at the time a big draw of the VCR was the ability to record TV programs. Even though the Magnavision cost less than a VCR, a consumer would be required to buy one in addition to a VCR. At the end of the day, the Magnavision offered higher-quality images, better sound, and features not yet seen on the VCR. But, for most consumers, that was not enough to convince them to purchase one instead of, or in addition to, a VCR.

WHY I LOVE IT

Being the first consumer videodisc player is certainly enough of a reason to add this to the collection, but I think the design is impressively futuristic. Magnavox was certainly pushing the design limits to make a modern piece of technology that looked and felt like it would have fit in on the set of *Star Trek*. This machine was truly ahead of its time.

HALLOWEEN II
DEBRA HILL _ JOHN CARPENTER · IRWIN YABLANS _ JOSEPH WOLF
A STANLEY KUBRICK FILM
JACK NICHOLSON · SHELLEY DUVALL
THE SHINI
MAGNAVOX
POWER
STILL
SLOW MOTION
PLAY
FAST
AUDIO
SEARCH
INDEX

MAGNAVOX
AUDIO
SEARCH
INDEX

1980 PIONEER VP-1000

Format: LaserDisc **Weight:** 38lb 8oz (17.5kg) **Introductory price:** $749 USD

Following the futuristic and extravagant styling of the Magnavision (see page 128), Pioneer brought the laser videodisc format back down to earth with a much simpler design that combined wood grain with lasers. Pioneer improved Magnavox's LaserDisc base by incorporating an infrared remote control and allowing for better manipulation of the picture-control capabilities inherent to the format. It also introduced—and trademarked—the term "LaserDisc," which became the commonly used term when referring to this type of media.

The VP-1000 could play both CAV and CLV discs. CAV stands for "constant angular velocity," meaning the discs spin at a constant speed of 1800 rpm, while CLV stands for "constant linear velocity," meaning the disc slows down from 1800 to 600 rpm as the laser moves from the center of the disc toward its edge. CAV discs are the "standard play" kind that only allow for up to 30 minutes of content per side, so an hour total per disc. Every circular "groove" contains just one frame of a movie. Therefore, as you get to the outer edge of the disc, a single frame occupies much more space than those at the center of the disc. These discs took advantage of all the features the format offers, such as freeze-frame, slow motion, and fast play. The CLV discs allowed for "extended play," and could hold up to one hour per side, for a total of two hours. They did this by adding more frames per "groove" as the "grooves" got larger. However, this prevented many of the features that made the format exciting and innovative from functioning.

As we saw with the Magnavision, the real advantage with the LaserDisc was the improved picture and sound quality, which made the format enticing for videophiles. LaserDiscs were capable of stereo sound at a time when it was not yet available on home VCRs. They also had a horizontal resolution of 340 lines, which was significantly higher than either VHS or Beta. The prices of players were still considerably lower than that of the VCRs at the time. The VP-1000 had a price tag of $749, when the average VCR was still $1,000 or more, and the discs went for about $25, while movies on cassette were priced between $50 and $100. As already discussed, one big drawback of the format was the inability to record television programs, but another situation that was muddying the waters was the prospect of other videodisc formats on the horizons. One of them was due to come from the technological powerhouse of RCA—the CED videodisc.

WHY I LOVE IT

I believe this is the only LaserDisc player that had a wood veneer, and I absolutely love that. It is like combining the past with the future, or low-tech with high-tech.

LASER DISC
PIONEER
LASER DISC PLAYER

LASER DISC
PIONEER

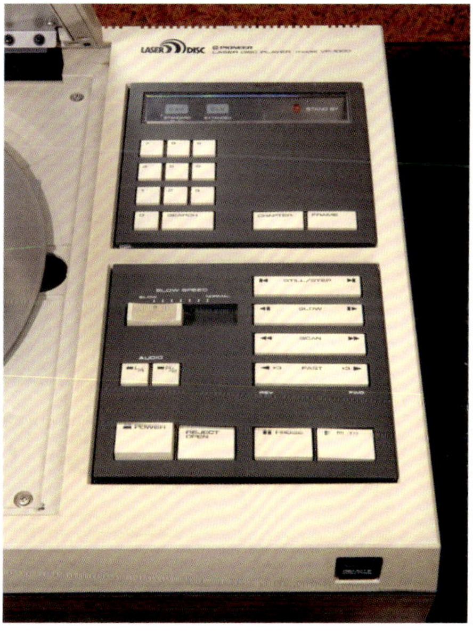
LASER DISC
PIONEER
STAND BY
SEARCH
CHAPTER
FRAME
SLOW SPEED
STILL/STEP
SLOW
SCAN
AUDIO
FAST
REV
FWD
POWER
REJECT OPEN
PAUSE
PLAY

1981 RCA SFT-100

Format: CED **Weight:** 20lb (9kg) **Introductory price:** $499.95 USD

And so, enter RCA with the CED videodisc. You have to keep in mind that when it came to home entertainment, RCA was top of the list for technological breakthroughs. As previously discussed (see page 13), it brought the TV into the home, developed the standard for color TV, and even played a significant role in the success of VHS. RCA was not to be taken lightly, and after 17 years of development, its CED format was finally ready to releas. Or was it?

CED was at a disadvantage from the start. Unlike LaserDisc, which used a laser to read the analog information on a disc, RCA's format used a stylus that physically interacted with grooves on a disc, not unlike playing a vinyl LP on a turntable. This physical interaction results in wear and tear, which limits the number of plays possible from the device and the disc. LaserDisc also had stereo sound from the start, but CED was introduced with just mono audio. Sure, most TVs were still mono at the time, but LaserDisc was being more forward-thinking and catering to the segment of the population interested in hooking up a player to their own sound system.

The design of the SFT-100 is immensely different from the design of LaserDisc players. While LaserDisc players were top-load machines, CED players were front load. They had to be, because the highly sensitive discs came built into plastic caddies to prevent them from being touched. These caddies made the format unique for years to come. In order to play a movie, the user has to insert the entire caddy into the player. The player then unlocks and removes the disc as the user pulls the caddy out. Although, at 240 lines,

RCA
SelectaVision | VideoDisc Player
RCA
SelectaVision
CED VideoDisc Player SFT 100

the horizontal resolution of the CED is similar to that of the VHS, when the technology is working properly the picture quality of the CED is noticeably better than VHS. Even in the early days of the formats, little difference could be seen between the image quality of CED and LaserDisc. While the CED caddy remains a fun novelty when compared to the advanced LaserDisc format, it already seemed like an outdated technology when it was introduced.

According to RCA, it was targeting "the average family"—people who were interested in a more affordable device for watching movie releases at home—so CED was not exactly in competition with the laser format. However, as both formats were videodiscs, they were pitted against one another. While competition can be good thing, it can also be confusing. Many video consumers had just experienced the rise and fall of several videocassette formats, so why would they want to get stuck backing a format that could quickly fail? It seems that this question was on the minds of many would-be buyers who thought it best to sit back and wait for a clear winner before investing.

"Compared to LaserDisc, the CED videodisc format was at a disadvantage from the start."

WHY I LOVE IT

In my opinion, CED is the most bizarre home-video format. The fact the discs are preserved in a plastic caddy that can only be removed by the player is unusual but interesting. And the way they seem to function in a similar way to a vinyl record while also producing a video images is surprising. At the time of its release, CED seemed to be using outdated tech while also being a new technological advancement. Although finding disc/player combinations that do not produce skips and other errors can be a struggle, this is a fun format to experience in person.

THE GOONIES
FAMILY CLASSIC
RCA
SelectaVision | VideoDisc Player
SIDE
PLAY TIME
RAPID ACCESS
VISUAL SEARCH
PAUSE
REV
FWD
REV
FWD

SIDE
PLAY TIME
RAPID ACCESS
VISUAL SEARCH
PAUSE
REV
FWD
REV
FWD
OFF
PLAY
LOAD | UNLOAD

ITEMS OF INTEREST 10

1 **CHANGE OF SCENERY**

Although the vintage electronics reside in my basement, the rest of the house also has a nostalgic vibe, with furniture and design cues taken from the Sixties and Seventies. From time to time, it is fun to display some of the electronics in other parts of my home.

2 **1976 SONY TRINITRON KV-1723D**

This 17-in Trinitron has a "high-rise" design in which the screen is placed above the controls. You may also notice that the set is not as deep as previous models. Sony had incorporated a wide-angle lens that allowed for a shorter tube and gave the TV a slimmer look.

3 **1978 SONY BETAMAX SL-8200**

The SL-8200 (see page 94) represents the transition between recording speeds, when recording speed doubled and image quality diminished as Betamax attempted to compete with the recording capacity of VHS.

1
2
3
SONY
L-500
L-750

1981 SONY LDP-1000

Format: LaserDisc **Weight:** 43lb 3oz (19.6kg) **Introductory price:** about $3,000 USD

Long before Sony started producing VHS VCRs, the company made LaserDisc players. Sony saw value in the format, but primarily for the industrial and educational markets. The Sony LDP-1000 is a professional-level videodisc player intended less for watching movies and more for educational and informational purposes. Some of these machines had Ford badges and were used at car dealerships for training and to demonstrate products to the customer. I also have one from an Ivy League university, Dartmouth College, that was presumably used in the classroom.

Like other consumer-level LaserDisc players of the time, the Sony is a top-load machine, but it is considerably larger and more easily serviced than others. A unique feature is that the controls located on the top-right corner of the player are set into a removeable, wireless remote control. This gives the user freedom to operate any of the player's functions from across the room. When no longer needed, the remote easily slides and locks back into place.

If having two different videodisc formats was not confusing enough, JVC (yes, the same JVC that was having great success with VHS) announced plans to release a third videodisc format. It was called Video High Density (VHD), and, in a similar vein to CED players, had a stylus, preserved the disc in a caddy, and made physical contact with the disc during playback. However, these two formats were not compatible. The VHD disc did not have grooves like the CED disc. Instead, it featured micro-pits that held the video and audio signal. VHD had been announced back in 1978, but it was several more years before the technology was ready to come onto the market. By this time, both LaserDisc and CED were struggling to stay afloat, and it was unlikely that VHD would do much better. Although it was on sale for a time in Japan and the UK, the VHD never made it to market in North America.

WHY I LOVE IT

Sony offered the highest-end players available at the time, and was not yet interested in lowering the price and making them accessible to the wider public. These machines are very cool, and the removeable remote is such a nice feature for a piece of analog technology.

THE SPECIAL EDITION
CLOSE ENCOUNTERS
OF THE THIRD KIND
TWIN PEAKS
SONY
SONY

MARK SHOSTROM ANTHONY SHOWE ED FERRELL
MICHAEL MINARD
JOHN DOUGLASS
NEIL WHITFORD
V35
LASER TUBE
AC OUT 120V 60Hz
MAX 400W
UNSWITCHED
AC IN
VHF
IN
VHF
OUT
CH 3
CH 4
FCC ID: AK896ALDP-1000A
CERTIFIED ONLY TO CE CODE
CERTIFIE EN VERTU DU CCE
SEULEMENT
VIDEO OUT
TV
EXT CPU
OFF ON
SC/SYNC
INT EXT
CENTERING
FOCUS
SC IN
SYNC IN
TRACKING
INDEX POSITION
LINE OUT
CH-2/R
CH-1/L
LOCK PULSE

1982 RCA SGT-200

Format: CED **Weight:** 21lb (9.5kg) **Introductory price:** $399.95 USD

By 1982, the CED had caught up in the audio department and the first stereo models were about to be introduced. The SGT-200 was the lower-end player at $399.95, and the SGT-250 was the more advanced version that cost about $50 more. The main difference between the two was that the SGT-250 was solenoid operated and had push buttons, whereas the SGT-200 still had a lever to load and unload discs. The SGT-250 came with a wireless remote and did not have the wood veneer used on the SGT-200. Although a remote control would be nice, I prefer the lever and the wood grain.

Another big improvement over the first-generation CED player was the addition of RCA video and stereo outputs. No longer was RF your only option. A switch next to the stereo outputs on the back of the machine enabled you to listen to the two audio tracks together, or isolate one of them if needed. This was useful if you had a bilingual disc and wanted to listen to your preferred language. With stereo players now available, stereo discs also had to be released. While all the early CEDs were monaural, stereo versions soon entered the market.

To help users identify the audio type of any particular disc, the mono ones were primarily supplied in white caddies, while the stereo discs were in blue caddies. This helpful system went out of the window when RCA ditched the CED and began putting new discs in whichever type of caddy was available.

Looking at the CED player, you will notice that it has significantly fewer buttons than a LaserDisc player or a VCR. While that may be a drawback if you are in the market for a ton of features, this is a simple machine that gets the job of playing a movie done with little to no confusion and head scratching. This was especially important when you consider that, at the time, more movies were available in the CED catalog than on LaserDisc, and many of the advanced LaserDisc functions were only available with the discs that stored 30 minutes of video per side.

WHY I LOVE IT

This machine offers an improvement over the audio quality of earlier CED players. At the time, stereo was a luxury not available with many formats. CED was never going to be as advanced as LaserDisc, but it was trying to entice customers looking for a more affordable disc player with features not yet found on videocassette formats.

RCA
SelectaVision
Stereo VideoDisc Player
STEREO
SIDE 1
SIDE 2
PLAY TIME
RAPID ACCESS
VISUAL SEARCH
PAUSE
OFF
PLAY
LOAD/UNLOAD

RCA
SelectaVision
Stereo VideoDisc Player
Capacitance Electronic Disc System

1983 PIONEER PR-8210

Format: LaserDisc **Weight:** 29lb 13oz (13.2kg) **Introductory price:** $1,500 USD

By 1983, both videodisc formats were established in the market, but that is not to say they dominated it. The sales of these players paled in comparison to VCR numbers. As both formats made improvements, it became clear that the image and sound quality were significantly better with LaserDisc players, especially when using composite connections. However, disc errors still plagued the laser format, and the chances of customers purchasing a defective disc were high. Fortunately, this issue would soon be addressed as Pioneer began taking over disc replication, which allowed for a higher number of discs to be produced with a lower number of defects. Pioneer was truly saving LaserDisc from obscurity.

This Pioneer PR-8210 was a higher-end player that was mainly intended for the industrial market. Some of the machines were even installed in arcade cabinets for interactive movies such as *Cliff Hanger* and *Dragon's Lair*. It is a very simple but elegant-looking pale beige player with a sleek black display across the front.

"In 1983, sales of videodisc players still paled in comparison to VCR numbers."

WHY I LOVE IT

This beautiful, top-load machine is one of my favorite LaserDisc players. It goes with almost any setup and can still look futuristic next to any television, whether it be CRT or flatscreen. This is my go-to player for watching LaserDiscs.

PIONEER
VIDEO DISC PLAYER PR-8210
LaserDisc
CONTROL KEY
SLOW SPEED
NORMAL 1/1
SLOW 1/30
PLAY PAUSE
SCAN
1/L
AUDIO
2/R
CX SYSTEM
REMOTE CONTROL SIGNAL

NIGHT OF THE CREEPS
PIONEER
LaserDisc
TOMY

1984 RCA SJT-400

Format: CED **Weight:** 21lb (9.5kg) **Introductory price:** $499.95 USD

The last CED players were produced in 1984, and RCA ended the run with the best player yet—the SJT-400. Not only had RCA fine-tuned the video and audio capabilities of the format, but the company also introduced interactive mode with this model. Using special interactive discs and the large 30-button remote control, the user could play interactive games in a choose-your-own-adventure style. For example, there was a murder mystery game that allowed you to make decisions by pressing buttons on the remote, which took advantage of the player's random-access functions. The SJT-400 was one of the few machines able to play these discs correctly.

By this time, RCA had sold close to half a million CED players and millions of discs. Even so, it was not enough to keep the format alive. In April 1984, RCA announced it would be discontinuing the production of players and discs, and, by the end of the following year, CED was gone. But it was not just a format that was lost; RCA was all-but lost as well. It had sunk over half a billion dollars into the failed format. The company that once played a big part in all facets of television, from production, to network stations, to the introduction of color broadcasting, was no longer capable of making those kinds of revolutionary advances. Shortly thereafter, RCA was bought by General Electric, and it never recovered from its downward slide.

CED was an astounding technology that came several years too late. There were already too many formats on the market that offered more than the CED could deliver. Although I love the CED format because of its unusual technology, I also think it's the most frustrating video format ever released. But in our modern times, where I can own and enjoy as many formats as I like, the quirks are what make CED something special. It may be difficult to find a disc that plays through completely without skips or errors, but inserting that plastic caddy into the front of the machine and then having the disc withdrawn instantly is an experience no other format can offer. It is the unusual aspects that make collecting analog equipment so interesting, and CED is nothing if not unusual.

WHY I LOVE IT

The random-access feature on CED was such great improvement, and it was surprising that RCA was actually able to make it work. Add to that the interactive discs, and this model of CED player is truly the most impressive of the format.

RANDOM ACCESS
RCA
SelectaVision Stereo VideoDisc Player

RCA 1984
VIDEO SYSTEMS
featuring RCA's new Digital Command Center
RCA
VIDEODISCS
DEMONSTRATION DISC

1993 PIONEER CLD-D702

Format: LaserDisc **Weight:** 19lb 6oz (8.8kg) **Introductory price:** $1,200 USD

Although the videodisc was on the verge of extinction in the Eighties, fast-forward to the early Nineties and you'll see a resurgence of the LaserDisc. During this time, the LaserDisc player evolved into a front-load machine, with a tray for you to place the disc on that extended from the player. Simply pressing a button would send the tray back into the machine for the disc to be loaded. Gone too were the days of special effects, such as freeze-frame and slow-motion viewing; these were reserved for CAV discs. Newer players incorporated digital memory that allowed these effects to be used with CLV discs as well. While the video remained analog, most LaserDiscs now used digital audio. In fact, the analog track was still there and could be utilized separately for things such as a special-feature commentary. A big downside to the early LaserDisc players was the inconvenience of having to flip the disc partway through a movie. This problem was also solved on the newer models thanks to the machine's ability to read both sides of the disc while it was in the player. Of course, if your movie was long enough to require multiple discs, you still had to get up to swap them.

The Shining by Stephen King
THE DEEP PETER BENCHLEY
STEPHEN KING CHRISTINE
VIKING
FROM THE TWILIGHT ZONE
Introducing CECILE BAGDADI
Produced by JOHN L. CHAMBLISS • MYRON MEISEL
RESTRICTED
UNDER 17 REQUIRES ACCOMPANYING PARENT OR ADULT GUARDIAN
AN MPM RELEASE
TWIN PEAKS
FIRE WALK WITH ME
"MESMERIZING"
"SHERYL LEE DESERVES AN OSCAR NOMINATION"
PIONEER
CLD-D703
BOTH SIDE PLAY
DIRECT CD
DISC SIDE
image ENTERTAINMENT
LaserDisc

Part of the reason for the resurgence of the videodisc can be attributed to the rise in popularity of the compact disc (CD) among audio enthusiasts. Many LaserDisc players were designed as combination or "combi" players, that could play LaserDiscs, CDs, and CD-Vs. The CD-V (Compact Disc Video) was a format that combined the audio CD with analog video. These 5-in discs could play five minutes of video in addition to the digital audio of a typical CD. The ability to play multiple high-quality formats ranging from music to movies on a single player was enticing to the many people who were looking for a superior entertainment experience. The LaserDisc player was no longer isolated to videophiles, it had now crossed over to an entire new demographic of audiophiles.

Another big plus for the LaserDisc in the eyes of the videophile was the introduction of widescreen versions of some favorite movies. Those in the know were aware of the fact that the traditional pan-and-scan of home video meant they were missing a lot of a movie. Theater screens have a wider aspect ratio than the TV at home, and so movies had to be cropped and edited to account for this change in format. LaserDisc finally brought the full cinema frame to your home. Although the average Joe complained about the black bars and the smaller image, widescreen was the next best thing to being in the cinema. Add to that the inclusion of special features that accompanied the movie, and the LaserDisc was really starting to shine.

The Pioneer CLD-D702 and CLD-D703 are currently the only newer LaserDisc players in my collection. They are both fine machines that take advantage of the technological improvements discussed above. Notice the CLD-D703 has a smaller tray that extends from the larger tray when a CD is being inserted.

"The resurgence of the videodisc in the early Nineties can in part be attributed to the rise in popularity of the CD among audio enthusiasts."

WHY I LOVE IT

The LaserDisc is an all-around impressive format that offers many unique features not seen on other formats during the early days of home video. While it did not win the overall format war against VHS, it was a stepping stone for soon-to-follow formats such as the DVD and Blu-Ray.

SOME MAY PASS THE TEST...
GOD HELP THE REST.
MPM Presents FINAL EXAM
Introducing CECILE BAGDADI • JOEL RICE
Executive Producers JOHN L. CHAMBLISS • LON J. KERR • MICHAEL MAHERN
Produced by JOHN L. CHAMBLISS • MYRON MEISEL • Music Composed by GARY SCOTT
Written and Directed by JIMMY HUSTON
PIONEER
CLD-D703
LaserDisc
no disc

BOTH SIDE PLAY
DIRECT CD
DISC SIDE
A
B
LaserDisc
V-DNR
STANDARD VARIABLE
HILITE /INTRO
REPEAT
V-DNR OFF
JOG & SHUTTLE
RANDOM PLAY
REV
FWD
Y-SIG
C-SIG
LASER DISC

FRIDAY THE 13TH
PART 3
3D
Stylin'
FRIDAY
THE 13 TM
PUSH START BUTTON

05

VIDEO CAMERAS, VIDEO GAMES & OTHER ACCESSORIES

The fun thing about collecting is that you never know which direction your collection will take you. While the primary focus of my collection is TVs and video players, I am always open to finding something new and interesting that fits in and stands out. I also feel it is necessary to incorporate more common items that help tell the story of the electronics from the Seventies, Eighties, and Nineties. You can't get a full picture of life in those times without also discussing the video games, video cameras, computers, and audio equipment that helped shape the fabric of our past.

Video games captured the imaginations of kids everywhere as they discovered there was more than one way to play. Video cameras made it more convenient to record special moments that may have previously just been preserved as a memory or photo. Computers expanded the limits of learning and communication that had been limited to books and telephones. Audio equipment allowed us to immerse ourselves in the artists and music that most spoke to our individualism. These technologies are just as much a reflection of the culture as the fashion, entertainment, and sports from the same era.

This section will highlight just a handful of additional electronics that have a comfortable spot in the basement. Some of them are rare and unusual, like the Combi 150 Jukebox that not only plays music but also plays films. Others are more common but still highly sought after, such as the Nintendo Entertainment System that was so ingrained in society that parents used the term "Nintendo" to refer to any game console the same way "Kleenex" is used to refer to tissues. So, let's pull a few of these items from the shelves and discuss how they fit into the basement, while appreciating the role they played in advancing us toward the technology we all enjoy today

1967 COLOR-SONICS COMBI 150 JUKEBOX

Format: 45 rpm vinyl record (audio), 8mm film cartridge (video)

This is an unusual piece in my collection. The Color-Sonics Combi 150 is a late-Sixties jukebox that not only plays vinyl records, but also films. I came upon this machine completely by chance while driving one day. I noticed it sitting on the porch of house that was being cleaned out, with everything going in the dumpster. I initially thought it was a cigarette vending machine and was surprised to learn it was a jukebox. Upon closer inspection, I noticed there was a "Film Selections" category, and quickly became intrigued. Knowing it was an uncommon machine to come across, I had to make sure it did not end up in the dump. After getting confirmation I could take it, I loaded it up and brought it to the basement.

The film options were not movies. They were some of the first music videos that could be enjoyed by patrons in a bar, which is where these machines were most often found. The films were originally shot in 35mm, which allowed for a higher-resolution image, and then duplicated onto 8mm film. This film was stored in a cartridge, which made it possible for the jukebox to automatically select and load the reel when the user made their choice. The film would then be projected straight up through a hole in the top of the jukebox, where a separate attachment housing a mirror and the projector screen was located. The image would reflect off the mirror and be projected onto the screen for viewing. Unfortunately for my machine, the top attachment with the projector screen had been taken to the dump before I was able to save it. Even so, this is a unique piece of entertainment history that deserved to be saved and will one day get the proper restoration it needs.

WHY I LOVE IT

The Combi 150 is a bizarre jukebox style that has practically been lost to history. There is very little information regarding these machines. I was not in the market for one or even knew they existed, but once I had discovered this one, I became intrigued, and I'm happy to have been able to save it.

color · sonics
combi · 150
film selections
record selections
quarters only

ITEMS OF INTEREST 11

1 **1970 SONY VIDEO CAMERA AVC-3400**

This camera would have been paired with the Sony Videocorder AV-3400 as a complete system called the VideoRover II, a portable, open-reel recorder that predated the videocassette. The camera was held in the hand while the recorder was strapped around the user's shoulder. It only captured images in black and white, but because this system did not use film that had to be developed, immediate playback of the recorded footage was possible. You could even watch the playback through the camera's viewfinder. It was close to a decade before a videocassette system with these functions would be available. I picked up this camera for cheap due to its poor condition, butI love the look of the "CABLE TV" decals that were applied to it at some point. I like to imagine this camera was used as a prop in some low-budget production in the vein of *Wayne's World*, because I doubt it was actually ever used by a cable TV channel. I do not own the recorder at the moment, but I would like to find a full example of the VideoRover II system with the camera in better condition. How cool would it be to walk around recording on reel-to-reel, while the videorecorder hangs from your shoulder?

CABLE
TV
SONY
SONY VIDEOCORDER
TRINITRON
SOLID STATE
VHF
UHF
1

1976 SONY AVC-1420 B/W VIDEO CAMERA

Format: Beta **Weight:** 2lb 2oz (0.99kg) **Introductory price:** $395 USD

In the early days of the VCR, a big draw to this new technology was the ability to make your own home movies and record them directly onto a videocassette. The Sony AVC-1420 was one of the first of these cameras, and it would have paired nicely with the Sony LV-1901 (see page 72) or another early Betamax VCR. At the time it could, however, be used with any VCR. The camera had no recording capabilities. Instead, it had to be linked to a VCR to capture the footage, which means the camera could only get as far from the VCR as the cable allowed. It also required a separate power supply. The camera was connected to the power supply, which was plugged into an AC outlet and output the video to the VCR. All the controls for recording had to be performed by the VCR, because the camera itself had no buttons or controls. To record sound, a separate microphone had to be plugged into the mic input on the VCR.

Although the ease of capturing video that could be played back immediately was a big improvement over earlier film cameras, which required time for the film to be developed, this infant technology still had a long way to go. It could only capture black-and-white images, required messy cables, and needed an abundance of light to perform at its best. Since the need for cables pretty much isolated it to indoor use, adequate lighting could be a challenge. But soon the home video camera would evolve into a revolutionary technology that would take the world by storm.

WHY I LOVE IT

This camera goes perfectly with my Sony LV-1901. There is even a port on the side of the console to hook the camera up and record through the Betamax.

1976 SONY AVC-1420 B/W VIDEO CAMERA

1979 APPLE II PLUS COMPUTER

Weight: 11lb 8oz (5.3kg) **Introductory price:** $1,195 USD

The Apple II computer was released in 1977 and quickly became popular, thanks to its color graphics and ease of use. It enjoyed widespread take-up in the education sector. The Apple II Plus was the second model released in the Apple II line. This model came with AppleSoft BASIC, which was a deluxe version of the BASIC language that the original Apple II came with.

I found this computer in a closet at an estate sale. It was still working and had all the original programs and accessories. It has a green-phosphor monochrome Apple Monitor III. These have a fine mesh across the front of the screen which was designed to reduce glare, and this can be easily damaged if not handled with care.

I also have a 1983 Apple IIe computer, which was the most popular model in the Apple II line. This "enhanced" model offered the ability to use lowercase letters, which had not been possible with previous models, and came with 64K RAM as standard. I have connected this computer to a white-phosphor monochrome Apple Monitor III.

The simple design of these computers, and the monochrome monitors, really capture that feeling of a new technology that was about to transform the world. In the following years, every aspect of our lives would be changed as the home computer flourished, and we all became connected by the internet.

WHY I LOVE IT

The style and simplicity of these early home computers give them a timeless look, while also showing how insanely advanced computers have become in such a short time.

Sargon II
for the best in
Apple II Chess
monitor ///
MICROSOFT SOFTCARD
MICROSOFT
SOFTCARD.
MICROSOFT
SOFTCARD.
MICROSOFT
apple II
APPLE II
disk II

monitor ///

1980 ATARI VIDEO COMPUTER SYSTEM

Format: Read-only memory (ROM) cartridge **Weight:** 3lb (1.3kg) **Introductory price:** $189.95 USD

I could not go through an entire book about vintage video without looking at a few video-game systems. The first home systems were introduced in 1972, and provided a new form of entertainment to play on the family television set.

The Atari Video Computer System (VCS) was released in 1977, and although it was not the first home video-game console on the market, it was one of the first to truly popularize this new pastime. It was originally designed with six switches across the top front, but a few years later two were moved to the back of the console and the number on the front was reduced to four. These two rear switches allowed the player to change the difficulty setting for each controller.

One of the things that made the Atari so popular was the option of playing a variety of different games stored on cartridges. Some earlier game systems featured just a single game, such as *Pong*, which was based on table tennis. These games provided hours of pleasure, but eventually the player lost interest and the fun diminished. Having a game console that allowed the user to play some of the best arcade games—as well as new games—in the comfort of their living room was a step in the right direction. When Atari released the 5600 in 1982, the VCS become better known as the Atari 2600. The Atari 2600 continued to be produced until 1992!

"A game console that allowed the user to play some of the best arcade games in the comfort of their living room was a step in the right direction."

WHY I LOVE IT

The Atari is an iconic game console that began many people's love of video games. It was one of the first consoles I remember encountering when growing up.

word zapper
PELÉ'S SOCCER
BERZERK
ASTEROIDS
COMBAT
ATARI
VIDEO COMPUTER SYSTEM
GAME SELECT
GAME RESET
POWER
ON
OFF
TV TYPE
COLOR
B•W

DIG DUG
CENTIPEDE
TENNIS
word zapper
PELÉ'S SOCCER
BERZERK
COSMIC ARK
PAC-MAN
ATARI
SLOT RACERS
ASTEROIDS
COMBAT
STAR RAID
VIDEO COMPUTER SYSTEM
GAME SELECT
GAME RESET
POWER
ON
OFF
TV TYPE
COLOR
B•W

1982 JVC HR-2650U VIDEO CAMERA SYSTEM

Format: VHS **Weight:** VCR: 12lb 8oz (5.8kg), camera: 6lb (2.7kg)
Introductory price: $1,100 (camera), $1,150 (VCR), $375 (tuner/timer)

This portable VCR and camera system was top-of-the-line in 1982. Essentially, it is JVC's table-top VCR split into two pieces: the VCR and the tuner. It has a wireless remote control that can be used for the VCR, or it can be wired to plug in and operate the video camera. The long-awaited feature of stereo audio is also found in this VCR. It had taken several years for VHS to achieve stereo audio. To do this, the one-millimeter audio track was split into two tracks separated by a guard band. This was a different technique than that used by Beta hi-fi systems. The JVC GX-S9U was the first stereo-ready video camera available in the US, and the HR-2650U followed suit.

Unlike the wired setup of the Sony AVC-1420 (see page 158), the JVC system allowed the user to take the VCR with them wherever they went. No longer were you tethered to the living room—you could just throw the battery-powered VCR over your shoulder and record on the go. Sure, you may have had a sore back by the end of the day, but you would always have all those home movies to look back at.

This particular system was originally purchased by my father-in-law in 1981. He dug it out of the closet and gave it to me knowing it would be well taken care of in the basement.

WHY I LOVE IT

While this multipiece video camera system is cumbersome, I think it looks much cooler than the bulky, all-in-one VHS camcorders that came later. Having the VCR strapped to your shoulder let everyone know you meant business.

JVC
IRIS
AUTO
FIX
MANU
CLOSE
OPEN
AUTO FADE
F.OUT
F.IN
C.OUT
PRESET
MANU.
WHITE
BAL.ADJ.
VIDEO
AUDIO
VIDEO
WHITE
BLACK
SENSITIVITY
UP
STD

JVC
REMOTE
COUNTER
RESET
MEMORY
TAPE RUN
INSERT
AUDIO DUB
REW
PLAY
FF
REC
STOP
PAUSE
POWER
REC LOCK
EJECT
PULL OPEN
CAMERA
REMOTE
JVC
VIDEO CASSETTE RECORDER
CHANNEL
TIMER
ON
OFF
CHARGE
JVC
TUNER ADAPTER

1983 SONY BETAMOVIE

Format: Beta **Weight:** 6lb (2.7kg) **Introductory price:** $1,499.95 USD

At a time when video cameras like the JVC HR-2650U (see page 164) consisted of several cumbersome parts, the Betamovie BMC-110 was the first video camera that combined the recorder, camera, and power supply in a single package. This new device became known simply as the "camcorder."

The Betamovie only recorded at the BII speed and used the standard Beta cassette, allowing it to record for up to 200 minutes on an L-830 tape. It was a rather basic camcorder, with record functions only. It was not capable of playback because its smaller head drum was designed to record a video signal that could only be played by a standard Beta VCR. Although you could not use the Betamovie to view what you had recorded, the fact that it was the first true camcorder on the consumer market attracted a lot of attention. According to Sony, in 1983, the Betamovie was selling faster than it could be produced. VHS companies had not yet figured out how to design a VHS camcorder, primarily due to the larger size of VHS cassettes, but a new type of VHS was about to change that.

"According to Sony, in 1983, the Betamovie was selling faster than it could be produced."

WHY I LOVE IT

Beta really should get a lot more credit for producing the first camcorder. I repaired mine and rebuilt the battery, so now I can take it on road trips and continue to capture memories on my favorite format.

SONY
Auto Focus
Betamovie
SONY
SONY
AutoFocus
Beta
SONY
Betamovie

1984 JVC VIDEOMOVIE GR-C1U

Format: VHS-C **Weight:** 4lb 5oz (2kg) **Introductory price:** $1,595 USD

The year after the Sony Betamovie became the world's first consumer camcorder, JVC released its own version—the GR-C1. Instead of designing a bigger camcorder to house a full-size VHS, JVC decided to shrink the cassette down to a more compact format. Unfortunately, these VHS-C cassettes only allowed for 20 minutes of recording. One improvement over the Betamovie, however, is the GR-C1's ability to perform all the typical functions of a VCR, including play, rewind, and fast-forward. You could also put the VHS-C into an adapter that made it function like a full-size VHS, allowing for playback on a home VCR.

The GR-C1 is most notable for its use in one of the greatest movies ever made—*Back to the Future* (1985). This is, of course, the camcorder Marty McFly uses to film Doc Brown at the Twin Pines Mall, before having to escape from terrorists and time travel back to 1955. Luckily, because the VideoMovie had all the basic functions of a VCR, it was possible for the 1955 Doc to hook it up to his TV set and watch the recording. He wouldn't have been able to do that with the Betamovie. A more recent sighting of this camcorder in pop culture was in season two of *Stranger Things* (2017), when the kids are trick-or-treating in 1984. In my opinion, it's the classic red accents and the sleek design that make this camcorder so iconic. It may be the best-looking one ever made.

"The GR-C1 is most notable for its use in one of the greatest movies ever made—*Back to the Future*."

WHY I LOVE IT

Back to the Future has long been my favorite movie and, as a tech guru, I had to own the same camcorder as Doc Brown. This thing is iconic!

VideoMovie
JVC
PLAY
PAUSE
STOP
EJECT
REC
REMOTE
EAR
JVC
TDK
TC-30

ZENITH
VAC 65 Cassette
TDK
TC-30
VHS
CASSETTE ADAPTER
ZENITH

ITEMS OF INTEREST 12

1 **SONY WALKMAN BROCHURE**

Many vintage electronics are not properly documented on the internet, so the best source of information can often be original dealer brochures and catalogs. I have binders full of this kind of paperwork in my collection.

2 **1983 SONY TRINITRON KV-1924**

This 19-in Trinitron from the early Eighties is another clean example of a moderately sized TV. By this time, Sony had advanced past tuning knobs, and each programmed channel had its own button in the column down the front of the console.

3 **1989 NINTENDO GAME BOY**

While Nintendo was dominating with the NES, it also released a handheld gaming system called Game Boy. It seemed like almost every kid had one of these in the Nineties. I remember dreaming about Tetris after playing it for hours at a time.

REALISTIC
Radio Shack
I NY
FRIDAY THE 13th
JASON TAKES MANHATTAN
SUPERIOR
OUTSTANDING
NOTABLE
YOUTHFUL
SONY
WALKMAN
SPORTS
1
2
3

1984 SONY STEREO SYSTEM

Format: Vinyl records, audio cassettes, compact discs

This Sony stereo system reflects the state of audio in the mid-Eighties. At the time, consumers often sustained an audio collection across multiple formats at once. As a new format came out, they did not completely abandon the old one, nor did they shun the new one and wait for it to fail. Typically they kept their old media and began buying new media for the more advanced format. With stereo systems such as this one, they could cover all the bases, and keep their collection moving forward.

The system consists of a turntable, cassette deck, AM-FM tuner, and amplifier. I added the CD player to it—it's an early model, the CDP-200. I also added the APM-615 speakers to complete it. I can listen to just about any audio format I have on this single tower.

I would have liked to have seen a similar technique for enjoying films on multiple formats utilized by videophiles, but in the early days of the home video there were too many companies trying to sell their own formats for this to work. It was not until the introduction of LaserDisc combi players, or the more popular VHS/DVD combination players, that this finally came to fruition. I have, however, made my own tower of video equipment to meet my needs.

WHY I LOVE IT

This is one of the best-looking stereo systems I've come across, and it meets all of my audio needs. The silver components stacked on the black stand just look perfect in the basement.

SONY
SONY
SONY
SONY
APM-615
PHONO

SONY
SPEED
ARM LIFTER
START/STOP
SONY
OPEN/CLOSE
POWER
HEADPHONES
LEVEL
COMPACT DISC PLAYER CDP-200 DIGITAL
PLAY
PAUSE
SONY
PHONO

1985 SONY CFD-5 BOOMBOX

Format: Audio cassettes, compact discs **Weight:** 17lb 7oz (8kg) **Introductory price:** $550 USD

As if the full Sony home stereo system (see page 172) wasn't enough, in the mid-Eighties there was also the boombox, which was capable of playing audio from several sources. The CFD-5 was the first boombox to be released with a CD player. At the time, the CD player was still in its infancy and did not react well to movement, so it was probably best to set this portable player down while playing a CD, and stick to the cassette player or the radio when walking around. Even so, having a CD player was a big deal back in those days, and having one that you could carry around and show off to people was even better.

Growing up, one of my favorite uses for the boombox was making mix tapes. Throwing a blank tape into the machine and waiting for my favorite songs to be played on the radio, so that I could record them, was a great lesson in delayed gratification. Once I finally had a full tape of songs, I would strap a smaller cassette player to the handlebars of my bike and my friend and I would cruise the neighborhood with the mix tape playing. That would pretty much be our entire summer. I doubt kids these days have anything as satisfying as that.

"In 1985, having a CD player was a big deal, and having one that you could carry around was even better."

WHY I LOVE IT

I always find the first models of a new technology to be particularly interesting, which is why I had to get the first boombox capable of playing CDs. The CFD-5 is a cool piece of audio history.

SONY
COMPACT DISC PLAYER
DIGITAL
SONY
AUTO REVERSE
DIGITAL

SONY
SONY
AUTO REVERSE
DIGITAL
SONY

1985 NINTENDO ENTERTAINMENT SYSTEM

Format: Read-only memory (ROM) cartridge **Weight:** 5lb (2.2kg) **Introductory price:** $99.99 USD

The Nintendo Entertainment System (NES) is possibly the greatest video game console ever released. It hit the North American market in 1985, just a couple of years after the video-game crash of 1983, at a time when many believed the video-games fad had already passed. This was not the case, and the NES quickly became a sought-after toy for children of all ages.

The 8-bit NES console brought an entirely new definition of fun to the family TV. Thanks to an abundance of software from third-party developers, there were plenty of games to keep users occupied. The Deluxe Set (and, later, the Action Set) came with the game *Duck Hunt*, along with the NES Zapper, which is my favorite accessory. Growing up, I didn't have an NES, but my best friend did, and we could often be found playing *Duck Hunt* on his small 12-in TV set.

Perhaps one of the reasons I love the look of the NES so much is because it was designed in the style of a front-load VCR. These machines were really hitting their stride around this time, and the NES designers took advantage of that, with a lift-up lid that allowed the cartridge to be hidden after insertion. With its black stripe and two-tone gray color scheme, this console has a timeless look. Almost anyone could recognize that iconic gray box with ease.

"The 8-bit NES console brought an entirely new definition of fun to the family TV."

WHY I LOVE IT

Who doesn't love the NES? To this day, people of all ages can pick up a controller and find enjoyment with this console. It's accessibility and simplicity has ensured its popularity persisted for many decades.

Nintendo
Nintendo
ENTERTAINMENT SYSTEM™
POWER
RESET
METAL GEAR
JAWS

Nintendo
ENTERTAINMENT SYSTEM™
POWER
RESET

Nintendo

1989 SEGA GENESIS

Format: Read-only memory (ROM) cartridge **Weight:** 4lb 4oz (1.9kg) **Introductory price:** $189 USD

"Genesis does what Nintendon't," was the slogan used to market the Sega system as a more advanced game console than the Nintendo. The 16-bit system offered a more arcade-like experience, with better graphics and a host of new games. One of them was Sonic the Hedgehog, which, in my opinion, is still one of the greatest video games ever made.

I had several friends who owned the NES, but none with a Sega Genesis, so this was the one I wanted. It was the first game console I ever owned and the one I grew up with. I remember my dad coming home from work after stopping at Sears and surprising my sister and me with the Genesis. It wasn't a birthday or special holiday; he just knew we wanted one..

We never had many games for it, but we spent many hours playing *Sonic the Hedgehog*. One other game we did have was *Ecco the Dolphin*, which I found very frustrating. Apparently, the goal is to search the ocean for the dolphin's missing pod. But all my sister and I did was swim around like a dolphin and talk to other creatures. We never successfully completed the game. We did, however, thoroughly enjoy *Sonic the Hedgehog 2*.

"The Sega Genesis offered a more arcade-like experience, with better graphics and a host of new games."

WHY I LOVE IT

I will forever love the Sega Genesis just because it was my first ever game console. That childhood memory means so much—not to mention the graphics were better than the NES.

NTAGE VIDEO
BASEMENT
• VHS •
• BETAMAX •
• LASERDISC •
• BLU-RAY •
PSYCHO
SONY
AUTO REVERSE
DIGITAL
PEPSI
Catch that Pepsi Spirit
PETER BENCHLEY JAWS
CHRISTINE
TRINITRON

06

WHEN PHYSICAL MEDIA WAS KING

A huge part of the basement is dedicated to physical media. The TVs, VCRs, videodisc players, video cameras, video game consoles, and every other device discussed in this book are all intended for the enjoyment of physical media. Physical media is not dead, but its Seventies, Eighties, and Nineties heyday has come and gone. That was the "Media Generation." Before then, people watched whatever programming was shown on the few TV channels they could pick up where they lived. There wasn't much choice. The media generation brought about the options and convenience of recording or purchasing the programs you were interested in. Always having access to your favorite movie, TV show, or video game meant owning a physical copy of it.

Nowadays, we have streaming services, which convince people that they have every option available to them all the time, but this is hardly the case. Today, buying a movie to watch is more likely to mean purchasing a digital copy. You may have bought a film, but that doesn't mean you own it. It can disappear or be altered without your knowledge or consent. It can also be hard to find a particular movie unless you subscribe to just about every service, and even then lots of things are not available to stream. Many programs are made specifically for streaming and the platform that produced them can decide to remove them at any time. If no physical copies of a show or film are distributed, it can basically disappear altogether. Therefore, physical media is still vital. VHS might have won the format war, but physical media is losing the content war against streaming. If we aren't careful, there will be no physical media left and the only thing we will own is a device that MAY allow us to access the content we cherish.

Today, people watch more shows and films than ever, but remember less. Content comes and goes so easily that we do not take it in or appreciate it. We do not even fully experience it, since we are usually scrolling on our phone while bingeing a full season of a show we will have forgotten by the end of the week. Trust me, I enjoy having a never-ending selection of new movies and documentaries to watch as much as the next person, but I believe taking the time to select and insert a physical copy of a film makes movie night much more meaningful. The fact that you had to put in a bit more effort increases the likelihood that you pay more attention and get more enjoyment from the experience. Delayed gratification is more satisfying than instant results because a sense of accomplishment comes along with it. Sure, talking about physical media with such earnestness may seem silly, but this ideology can be applied to other areas of our

"There is something intoxicating about owning your favorite shows, movies, video games, and home recordings as physical objects."

lives. The easy way is rarely the best way, and things worth having are worth working for. Repairing and using obsolete electronics is way of appreciating something that brings me happiness and enjoyment.

There is something intoxicating about owning your favorite shows, movies, video games, and home recordings as physical objects. Certainly, streaming makes it easier and more convenient to watch some content, but I am a firm believer that "easier" and "more convenient" are not necessarily synonyms for "better" or "more enjoyable." I'd much rather grab a videocassette, turn on the VCR, load the tape, and press play than wait for software updates and try to remember usernames and passwords. Don't get me wrong. I do have modern flatscreen TVs, high-speed internet, and access to many streaming platforms, but it's nice to do things the old-fashioned way occasionally, even if that involves a Blu-Ray disc. Luckily, nowadays I can enjoy all formats, from the lowest quality up to 4K UHD. I think they all offer a different type of pleasure and satisfaction. For me, physical media is still king.

MYTHS ABOUT THE DEATH OF BETA

We have now seen that VHS made a lot of good calls that lead to its success and dominance for home video, so let's get one thing straight—porn did not kill Beta. As we explored throughout the section on the war between Beta and VHS (see Chapter 3), many different factors played a role in VHS becoming the dominant format. In particular, it was cheaper, the tapes were capable of longer record times, and the equipment was more readily available. Consumers did not choose VHS because the porn industry had picked it, the porn industry picked VHS because consumers had already identified it as their preferred videocassette format. Video rental stores were also already phasing out their Beta purchases. When space is limited, a store can offer customers a much better selection if it doesn't have to stock the same film in multiple formats.

ZENITH
VHS
BETAMAX
LASERDISC
BLU-RAY
SONY
SOLID STATE
TRINITRON

PRESERVING THE PAST

Now that you have viewed my collection, learned about the history of home video, and seen amazing electronics from the analog age, you may be inspired to start your own basement or dedicated collection. So how do you get started? I would recommend either deciding on the particular types of electronics you like and then basing your collection around those, or choosing the time period you most enjoy and using that as your starting point. My collection started with VHS, and evolved into other formats before becoming a collection of video players and televisions that enhanced the experience of watching analog media. At the same time, I accumulated furniture, toys, and ephemera to decorate the space around the tech.

There is no right or wrong way to collect, but my preference is to stay within a certain aesthetic. I collect pretty much anything I deem to be cool, and everything complements the rest of the collection and fits within the same style. Therefore, collect what you like and incorporate your own style. That being said, I do feel there are certain rules that make almost any collection better and more appealing.

1 COLLECT YOUR PASSION

Preserve the things you care most about and not necessarily what could be worth most in the future. Some people look at collecting as an investment and save items primarily for their potential value. Although these collections may still be impressive, they are rarely meaningful. Others can tell when you truly have a passion for your collection. People often ask me what an item in my collection is worth, and I rarely have an answer. I know what I was willing to pay for something, but I don't often know what others would be willing to spend. I preserve these machines for historic and personal reasons, not for monetary value. While the prices of old CRTs and VCRs have increased rapidly recently, there are still free and cheap options to be found. Of course, it's nice if the things you are passionate about also happen to increase in value, but if your intention for starting a collection is to someday sell it all, did you really care about it to begin with?

REALISTIC
Radio Shack
1
MICHAEL J. FOX
CHRISTOPHER LLOYD
MARY STEENBURGEN
They've saved the best trip for last...
But this time they may have gone too far.
BACK TO THE FUTURE III
OPENS MAY 25TH
Stylin'
E.T.
QUEEN
SONY
BLOCKBUSTER VIDEO
BLOCKBUSTER

2 MAKE IT COZY

To ensure that your collection is something you and others want to enjoy, make the space inviting. If you intend it to be a place to hang out and lounge in, get appropriate seating. I like to use vintage furniture that fits the era and vibe of my collection, whereas others may prefer new pieces. For me, searching for the furniture is as much fun as searching for the electronics, but it is much easier to run to the store and pick out a new couch. Either way, make sure it will be comfortable for you and your guests.

3 SET THE MOOD WITH LIGHTING

To make your collection really stand out, pay close attention to the lighting. Do not just rely on overhead lights. As a matter of fact, do away with them as much as possible. Overhead lighting is not interesting or flattering. It is often stark and casts ugly shadows. Bring in other light sources such as floor lamps, desk lamps, and illuminated signs. Do not be afraid to use multiple lamps. I basically collect lamps—I have dozens of them spread throughout the basement—but they are necessary because some items look better backlit, while others look best front-lit. Arranging multiple lamps and light sources can give your display more depth and interest.

Another important aspect of lighting is the color temperature of the bulbs. Color temperatures below 3500K have a warmer, more inviting color. Above this temperature, the light is more white and begins to transition to blue. It tends to be less appealing to the eye and can feel sterile, like a hospital. Warmer lighting is a necessity if you want to create a cozier environment. A collection rarely needs stark, bright lighting, so unless you want your room to look like a dentist's office, go with the warmer tones of soft light.

4 POT UP SOME PLANTS

After you have set up your collection, incorporate some plants. This tip is almost always neglected, but it can breathe life into your room. For example, if my basement was just an assortment of electronics, it would feel cold and sterile. Adding a few houseplants, even if they are fake, makes it feel lived -in and more comfortable. Another advantage of houseplants is that you can use them to fill awkward spaces where nothing else really fits, and they help hide the many electrical cords that are routed around the room.

3
4
The
TWILIGHT
ZONE
SONY
FROM THE TWILIGHT ZONE
2

5

SCATTER A RUG

Another big improvement that can be made to your space is the addition of a rug or carpet. A hard floor in an area intended for lounging or comfort is not enjoyable—just putting down a rug will give the room a more relaxing feel. Throughout the Eighties and Nineties, I grew up in a home with brown carpet (and brown everything, really), so when I was designing the basement, I wanted to give it the feeling of what I experienced those times to be like. Instead of brown, I chose a dark orange rug for the primary area, and it really differentiated the basement from all the other collections I had seen. You would not believe how many times I get asked about that rug, it's as popular as most of the electronics!

6

DECORATE THE WALLS

Along the same line, dress up the walls. I was fortunate enough that our current house still had the original wood paneling throughout the basement when we bought it. That was a big plus. However, that was not the case for our previous home. I had to search salvage stores and renovated homes for a few pieces of wood paneling to hang. Everything was wood paneled when I was growing up, so it provides a sense of comfort. Finding good paneling can be difficult. If that is the case, pick an interesting paint color or use some textured wallpaper instead. Whatever you choose, be creative and make sure it fits the vibe you are going for.

7

SEARCH FOR STUFF

People often ask me where they can get a TV like mine, and I never know how to answer. It is not like there is a store that has them in stock. The real answer is that you have to be ready to hunt for it, and it will often be a long, intensive search. Some of my favorite places to search include Facebook Marketplace, Craigslist, eBay, estate sales, and yard sales. Of course, the available selection often depends on the area where you are located. There will be many more options in big cities, but the prices are usually higher, and you may have to compete with other collectors searching for similar items. If you live in a more rural area, good listings may be few and far between, but the prices are often lower due to there being less interest. No matter where you are searching, know what you like and be ready to grab it when you see it. If you are looking for really unusual pieces, you may have to be willing to do some traveling. If what you want isn't nearby, go to where it is. I have made several multi-day roadtrips in order to pick up some of my favorite items. The thrill of the hunt is often the best part of collecting.

A New Dimension In Terror...
FRIDAY THE 13TH
PART 3
3D
Stylin'
6
7
FRIDAY
THE 13TH
TM
PUSH START BUTTON
SONY
BLOCKBUSTER VIDEO
5

THANKS FOR VISITING

I hope you have enjoyed plundering through the basement and checking out many of the incredible electronics that came before our modern-day devices. Maybe you have also been inspired to start your own collection to help preserve these relics for future generations. At the very least I hope you understand these obsolete machines can still offer enjoyment and should be saved from landfill. I continuously scour the internet for unusual and often unwanted pieces. Even though I do not have everything, and likely never will, the thrill of the hunt and the desire to learn more make collecting such a powerful drug.

With the number of TVs, VCRs, and other devices that I already have, and the fact that I try to enjoy them all, there is a constant queue of maintenance and repair to be performed. That is another aspect of the hobby that I find enjoyable. There isn't much that feels better than bringing a once-dead device back to life. If you have any equipment that stops working, and you are not inclined to do the repairs yourself, please do not throw it out. Consider having it repaired or giving it to someone interested in doing so.

Many people are discovering or re-discovering the joys of using CRT televisions to play classic video games, watch VHS, and even make art. This technology, that was destined for the dump just a few short years ago, is enjoying a resurgence in popularity. It is beginning to pop up in antique stores, thrift stores, and online listings as people begin to see its value, and prices reflect that. Although the days of finding items for free on the side of the road seem to be over, with a little patience and determination, good stuff can still be found at low prices. The same goes for VCRs. People who collect vintage physical media such as VHS, Beta, and LaserDisc will continue to need players. We must save and repair them as much as possible. Here in the basement, it is not about living in the past. It is about preserving the parts of the past that are loved the most.

HALLOWEEN III
SEASON OF THE WITCH
The night no one com
ALLOWEEN II
The Nightmare Isn't Over!
HE NIGHT NO ONE
OMES HOME
HALLOWEEN III
SEASON OF THE WITCH
HALLOWEEN III
Season of the Witch

ABOUT THE AUTHOR

Matt Parker has been collecting, repairing, and preserving vintage electronics since childhood. In addition to electronics, he also collects physical media, toys, furniture, and anything else that fits his aesthetic and personal style. He uses social media to introduce obsolete technology, educate others about it, and evoke a bit of nostalgia for a simpler time.

Matt was raised in Cheraw, South Carolina, and obtained a master's degree in physics from the University of Georgia. Later, he became a debug engineer at a tech startup in Washington state. He now resides in Kingsport, Tennessee, with his wife, Lauren, and son, Charlie.

ACKNOWLEDGMENTS

The basement began as a way for me to display and enjoy the many things I had collected over the years. I must acknowledge that my wife supported my crazy accumulation of obsolete tech, and even assisted in locating and acquiring many of my favorite pieces. She also has her own collection of vintage items (which I borrow on occasion for videos and photo shoots). Our love of collecting allows us to spend time together going to estate sales, antique stores, and thrift stores. I wouldn't have such an amazing collection if it were not for her. Thanks, Lauren.

I would like to acknowledge my son Charlie, who turned two years old in the process of writing this book. Much of it was written with him in my lap or playing nearby, and I wouldn't want it any other way. He already has his own VHS collection and knows how to work the VCR and CRT in his room. I'm a proud dad!

Additionally, thanks to my editor Stephanie Selçuk-Frank for all the feedback, recommendations, and help to ensure this book is great and accurately represents the basement and my collection. I am very grateful for this opportunity. Thanks, Steph.